A HISTORY OF THE SCIENCE
AND POLITICS OF
CLIMATE CHANGE

The Role of the Intergovernmental Panel
on Climate Change

BERT BOLIN

University of Stockholm
IPCC Chairman 1988–1997

CAMBRIDGE
UNIVERSITY PRESS

CAMBRIDGE UNIVERSITY PRESS
Cambridge, New York, Melbourne, Madrid, Cape Town, Singapore, São Paulo, Delhi

Cambridge University Press
The Edinburgh Building, Cambridge CB2 8RU, UK

Published in the United States of America by Cambridge University Press, New York

www.cambridge.org
Information on this title: www.cambridge.org/9780521880824

First published 2007
This digitally printed version 2008

A catalogue record for this publication is available from the British Library

ISBN 978-0-521-88082-4 hardback
ISBN 978-0-521-08873-2 paperback

Contents

Foreword

Bo Kjellén

As a climate negotiator in the early 1990s I have a strong recollection of the impact of Professor Bolin's statements to the International Negotiating Committee for the Framework Convention on Climate Change. When the chairman of the Intergovernmental Panel on Climate Change (IPCC) presented its findings there was silence in the room: here were the facts, the certainties and the uncertainties.

We were all part of a process in which national interests and national instructions governed our actions and limited the rate of progress. We were all painfully aware of this, and we were also on a learning curve. As diplomats and generalists, most of us had limited knowledge of the substantial issues of climate change, but here we had the opportunity to listen to one of the most prestigious experts, speaking in clear language, devoid of academic jargon. Furthermore, we realised that Bert Bolin, as a former scientific adviser in the Swedish Prime Minister's office, had a thorough knowledge of the political process, its possibilities and limitations.

All this enabled him to set high standards for the work of the IPCC from the beginning, creating a scientific backstop to the negotiations which in my view has had a decisive impact on the relative success of the process. The IPCC is not only a venue for interdisciplinary science, it is also a meeting-place for researchers and Government officials, thereby facilitating the inevitable process of multilateral bargaining on the terms of legally binding international instruments.

As the discussions and negotiations for the climate regime after 2012 now get under way, it is of great importance that negotiators have a clear picture of the background to the negotiations, and that they realise the full importance of the subtle interaction between scientific research and progress in the negotiations. This book provides an inside view and an authoritative interpretation of the process which will no doubt assist in the difficult tasks ahead. It will also help all interested to get a clearer picture of the status of climate research and of the energy futures that will be decisive for global economic and political relations all through this century.

However, there are also wider issues involved. Changes in immense global systems brought about by human influence go beyond climate. Freshwater, oceans, desertification, fisheries and biodiversity are all issues that create serious threats for the future. We are only beginning to grasp the complicated systemic problems involved; still less do we understand how our society can best cope with them. But we realize that sound scientific research – within both the natural and the social sciences – is necessary to provide background for political action. The IPCC approach may provide important clues to how to tackle other global problems.

One final remark about the nature of these threats, and their impact on the international political system: in my view, the fact that we risk creating irreversible damage to the planet's life-supporting systems forces us to consider new objectives in international cooperation in order to ensure the welfare of future generations. Therefore I believe that a new diplomacy for sustainable development is emerging, still in the shadow of traditional diplomacy with its reliance on national security, ultimately through military means. As the character of global threats of a new kind is more clearly understood, it may well be that this new diplomacy will create different and better ways of dealing with common problems, opening new avenues for multilateral cooperation in the UN framework, at present clearly in crisis. Since this diplomacy for sustainable development is so dependent on scientific research, the IPCC story is worth considering very carefully.

Abbreviations

ACIA	Arctic Climate Impact Assessment (IASC)
AGBM	Ad-hoc Group on the Berlin Mandate (FCCC, 1995–1997)
AGGG	Advisory Group on Greenhouse Gases (ICSU/UNEP/WMO)
AMS	American Meteorological Society
AOSIS	Alliance of Small Island States
AR4	IPCC Assessment Report No 4. (2007)
BAS	British Antarctic Survey
CAS	Committee on Atmospheric Sciences (IUGG/ICSU)
CCS	Carbon Capture and Storage
CFC	Chlorofluorocarbons
COSPAR	Committee on Space Research (ICSU)
CSIRO	Commonwealth Scientific and Industrial Research Organisation (Australia)
FAR	First Assessment Report (IPCC, 1990)
FCCC	Framework Convention on Climate Change (UN)
FGGE	First GARP Global Experiment (JOC, 1978–80)
GARP	Global Atmospheric Research Programme (ICSU/WMO, 1967–1980)
GATE	GARP Atlantic Tropical Experiment (JOC, 1974)
GCM	Global Circulation Model
GPP	Gross Primary Production
IAMAP	International Association of Meteorology and Atmospheric Physics (IUGG)
IASC	International Arctic Science Committee
ICAO	International Civil Aviation Organisation
ICSU	International Council of Science (earlier; Scientific Unions)
IEA	International Energy Agency
IGBP	International Geosphere Biosphere Programme (ICSU)

IHD	International Hydrological Decade (UNESCO)
IIASA	International Institute for Applied Systems Analysis
INC	Intergovernmental Negotiating Committee (UN)
IOC	International Ocean Commission (UNESCO)
IUBS	International Union of Biological Sciences (ICSU)
IUCC	Information Unit on Climate Change (WMO/UNEP)
IUGG	International Union of Geodesy and Geophysics (ICSU)
JOC	Joint Organising Committee of GARP (ICSU/WMO, 1968–1980)
LCA	Life Cycle Analysis
MAB	Man and the Biosphere (UNESCO)
MIT	Massachusetts Institute of Technology, Cambridge, MA, USA
NAS	National Academy of Science, USA
NASA	National Aeronautics and Space Agency, USA
NBP	Net Biome Production
NCAR	National Corporation for Atmospheric Research, USA
NEP	Net Ecosystem Production
OECD	Organisation for Economic Cooperation and Development
ppmv	parts per million of volume
SAR	Second Assessment Report (IPCC, 1995)
SBI	Subsidiary Body for Implementation (FCCC)
SBSTA	Subsidiary Body for Scientific and Technolgical Advice (FCCC)
SCEP	Studies of Critical Environmental Problems
SCOPE	Scientific Committee on Problems of the Environment (ICSU)
SMIC	Study of Man's Impact on Climate
SRES	Special Report on Emission Scenarios (IPCC, 2000)
TAR	Third Assessment Report (IPCC, 2001).
TEAP	Technology and Economic Assessment Panel (Parties to the Montreal Protocol)
TERI	TATA Energy Research Institute (Bombay)
UCAR	University Corporation for Atmospheric Research, Boulder, CO, USA
UGGI	Union de Geodesie et Geophysique International (see IUGG)
UNCED	United Nations Conference on Environment and Development (Rio, 1992)
UNEP	United Nations Environmental Program (UN)
UNESCO	United Nations Educational, Scientific and Cultural Organisation
URSI	Union Radio Scientifique International (ICSU)

UTAM   Union of Theoretical and Applied Mechanics (ICSU)
WCED   World Commission on Environment and Development,
           Brundtland Commission (UN, 1984–1987)
WCRP   World Climate Research Programme (WMO/ICSU/IOC, 1980)
WMO   World Meteorological Organisation
WWW   World Weather Watch (WMO)

Part One

The early history of the climate change issue

1

Nineteenth-century discoveries

Variations of atmospheric concentrations of carbon dioxide may well change the global climate.

The nineteenth century saw a remarkable development of our knowledge about past climatic variation. The French natural philosopher Joseph Fourier (1824) put forward the idea that the climate on earth was determined by the heat balance between incoming solar radiation ('light heat') and outgoing radiation ('dark heat') and this idea was further pursued by Claude Pouillet (1837). They both realised that the atmosphere might serve as an absorbing layer for the outgoing radiation to space and that the temperature at the earth's surface might therefore be significantly higher than would otherwise be the case.

At about the same time the Swiss 'naturalist', Louis Agassiz (1840) suggested that features in the countryside, such as misplaced boulders, grooved and polished rocks, etc., were indications of glacial movements and that major parts of central Europe, perhaps even northerly latitudes in general, had been glaciated. This revolutionary idea was, of course, not readily accepted by his colleagues, but it stimulated others to search for further evidence. Agassiz's idea found acceptance during the following decades, not least because of his studies in the Great Lakes area in the USA.

The idea that the atmosphere plays an important role in determining the prevailing climate of the earth was further developed in England by John Tyndall (1865). He actually measured the heat absorption of gases, including carbon dioxide and water vapour, and emphasised their importance for the maintenance of the prevailing climate on earth. He thought that variations of their concentrations might explain a significant part of the climate variations in the past. Thus Tyndall clarified qualitatively what we today call the *greenhouse effect*, but he did not attempt to quantify its role. Data were simply inadequate to do so.

3

Agassiz's discoveries and work by other researchers in central Europe also attracted geologists in Scandinavia, particularly Gerhard De Geer in Sweden, who contributed greatly to the advance of our knowledge of glaciations over Scandinavia. De Geer studied the layers of clay that can be found in lakes and in areas earlier submerged by lakes or by the sea at the time of the decline of the major ice sheet over Scandinavia. He was able to show that the layers represent annual deposits of particles that were set free in the course of melting and carried by the runoff of the melt water to less turbulent places where deposition could occur. He was able to use his extensive data set to determine accurately the chronology of the withdrawal of the Scandinavian ice sheet.

The natural questions to ask were of course: Why did the climate become warmer some 10 000 years ago? How long had there been an ice age? Obviously the heat balance between the earth and space must have been disturbed in some way. It was already known at that time that the elipticity of the earth's orbit around the sun varies regularly, which creates a periodic variation of the incoming solar radiation and its distribution over the earth. James Croll in England considered such variations as the most likely reason for the observed variations of climate. Alternatively, the optical characteristics of the atmosphere or the earth's surface might have changed, but why?

This was the state of knowledge in the early 1890s when a group of scientists at Stockholm's Högskola[1] addressed the issue anew under the leadership of Svante Arrhenius.[2] He had recently been appointed teacher of physics at the Högskola and was keen for his research to be of relevance to society. He had put the physics of our environment in the broad sense of the word high on his agenda. To some extent this was a protest against the traditional role of many universities in the late nineteenth century, particularly the University of Uppsala as Arrhenius had experienced himself. He had had great difficulty in having his doctor's thesis approved at Uppsala some ten years earlier, but since then had gained international recognition for his development of the theory of the dissociation of solutions. The relations between the faculties in Stockholm and Uppsala remained tense.[3]

Under Arrhenius' leadership some remarkable discussions and analyses were initiated. As one of his first actions as professor at Stockholm's Högskola he founded the Stockholm Physics Society. The members met every other Saturday morning for a public seminar. Lectures were given and the discussions were open and lively. The group included: Vilhelm Bjerknes, professor of theoretical physics, later renowned for his development of physical hydrodynamics, who thus provided a solid foundation for modern meteorology; Otto Petterson, oceanographer; Arvid Högbom, geologist and one of the first to analyse the circulation of carbon in nature; and Nils Ekholm from the Swedish Meteorological Office, a specialist in atmospheric radiation.

Arrhenius' decision in 1894 to study the mechanisms of climate change was probably a result of a presentation by Ekholm of Croll's idea that climate variations were primarily caused by variations of solar radiation and another one by Högbom describing sources and sinks for the carbon dioxide in the atmosphere, both given as Saturday seminars. Arrhenius wanted to determine the sensitivity of the climate system to changes of the water vapour and carbon dioxide concentrations in the atmosphere. He was intuitively sceptical of Croll's view about the importance of variations of solar radiation and was curious about the magnitude of possible variations of the greenhouse effect due to changes in the concentrations of water vapour and carbon dioxide in the atmosphere. However, this required knowledge of their radiative characteristics. Adequate laboratory measurements were not available, but the American physicist Langley (1889) had deduced the temperature of the moon by observing its dark (infrared) emissions. Arrhenius realised that these data could also be used to determine quantitatively the absorption by the atmosphere due to the presence of these heat-absorbing gases by evaluating the intensity of their absorption as a function of the angle of elevation of the moon.

Arrhenius also recognised early that there is a most important feedback mechanism that must be considered. If the air becomes warmer because of an increasing carbon dioxide concentration, it is likely that the amount of water vapour in the atmosphere will also increase because of enhanced evaporation. This would in turn cause additional warming. Conversely, cooling would be enhanced if the carbon dioxide concentration were to decrease. In fact, the plausible assumption made by Arrhenius that the relative humidity probably would remain unchanged yields an enhancement of the warming due to an increase of the carbon dioxide concentration of at least 50%. It is interesting to note in passing that the magnitude of this feedback mechanism was a controversial issue until the 1990s. Let us recall Svante Arrhenius' own description of the greenhouse effect as given in a popular lecture early in 1896:[4]

As early as at the beginning of this century, the great French physicists Fourier and Pouillet had established a theory according to which the atmosphere acts extremely favourably for raising the temperature of the earth's surface. They suggested that the atmosphere functioned like the glass in the frame of a hotbed. Let us suppose that this glass has the property of transmitting the sun's rays so that objects under the glass are warmed, but not of transmitting the heat radiation emitted by the object under the glass. The glass would then act as a sort of trap which lets in the heat of the sun but does not let it out again, when it has been transformed to the radiation of bodies with a lower temperature. Glass does in fact act in this way, as has been shown by experiments, although only partially, not totally, so. According to Fourier and Pouillet a similar role is played by the earth's atmosphere which, one might say, retains the sun's heat for the earth's surface. The more transparent the air becomes for the sun's rays, and the less it becomes so for the heat radiation from the earth's surface, the better it is for the temperature of the earth's surface.

The transparency of the air depends principally on three factors. Extremely fine suspended particles in the air impede the penetration of the sun's heat, although they have little effect on the heat radiated by the earth. Further, the clouds reflect a great deal of the sun's heat which impinges on them. The main components of the air, oxygen and nitrogen, do not absorb heat to any appreciable extent, however, the opposite is true to a high degree for aqueous vapour and carbonic acid in the air, although they are present in very small quantities. And these substances have the peculiarity that to a great extent they absorb the heat radiated by the earth's surface, while they have little effect on the incoming heat from the sun.

It should be pointed out, however, that the analogy of the hotbed (or, as we say today, greenhouse) is deficient in one important way. The glass has an additional function in a greenhouse in that it prevents the hot air beneath it escaping. The atmosphere, on the other hand, is often mixed by convective currents, whereby heat is transferred to higher levels, from where radiation to space takes place. The term greenhouse effect has, however, come to stay, since it describes an important mechanism simply, though not perfectly.

Arrhenius spent most of 1895 carrying out the very tedious computations that were required to give a quantitative answer to the question he had asked. He kept the members of the Physics Society informed by giving two presentations in the course of the year. In 1896 his paper on this work was published by the Royal Swedish Academy (in German) and the *Philosophical Magazine* in England (Arrhenius, 1896a).

Arrhenius presented the expected change of the surface temperature as a function of latitude and time of the year for carbon dioxide concentrations equal to 0.67, 1.5, 2.0, 2.5, and 3.0 times the prevailing concentrations, which were assumed to be about 300 parts per million of volume (ppmv). He thus explored the consequences of both a decrease and an increase of carbon dioxide concentrations. The spatial and temporal distributions that he determined are of secondary interest, since in reality the motion of the air would change these distributions, but he determined that the average global change of surface temperature due to a doubling of the carbon dioxide concentration would be $5.7\,^\circ$C. He recognised that the precise magnitude of the warming is uncertain and he later reduced this figure somewhat on the basis of additional computations.

Arrhenius drew the conclusion that variations of the amount of carbon dioxide in the atmosphere might well be an important factor in explaining climate variations thereby refuting Croll's hypothesis. He referred to the view expressed by Högbom that volcanic eruptions add carbon dioxide to the atmosphere, but there were no data to support his view that this might have been the reason for the ending of the last ice age.

Arrhenius also explored the possibility that human emissions of carbon dioxide might bring about a global warming. The annual emissions due to coal burning at that time were about 400 million tons of carbon, i.e. 0.7 per thousand of the amount present in the atmosphere. He believed that a significant part of these emissions must, however, be removed by the dissolution of carbon dioxide in the sea. He rightly pointed out that at equilibrium only about 15% would stay in the atmosphere but did not realise that the turnover of the sea is a slow process and that it actually takes more than a millennium to reach equilibrium. We know today that only about 20% of the emissions to the atmosphere since the beginning of the industrial revolution some 150 years ago have dissolved in the sea. However, Arrhenius did not know that the use of fossil fuels would increase very rapidly, in fact by a factor of about 15 during the twentieth century. He therefore dismissed the possibility that man one day might cause significant global warming, but would have welcomed such a development. He actually wrote (Arrhenius 1896a): 'It would allow our descendants, even if they only be those in a distant future, to live under a warmer sky and in a less harsh environment than we were granted.'

Arrhenius' evaluation of the greenhouse effect is a remarkable achievement. This is brought home by two leading researchers in the field today, Ramanathan and Vogelmann (1997), who characterise his work as follows:

Svante Arrhenius laid the foundation for the modern theory of the greenhouse effect and climate change. The paper is required reading for anyone attempting to model the greenhouse effect of the atmosphere and estimate the resulting temperature change. Arrhenius demonstrates how to build a radiation and energy balance model direct from observations. He was fortunate to have access to Langley's data, which are some of the best radiometric observations ever undertaken from the surface. The successes of Arrhenius model are many, even when judged by modern day data and computer simulations.

Arrhenius' analysis of the climate change issue was discussed for a few years, but there were not enough data to tell whether he was right or wrong. The amount of carbon dioxide could not be measured with sufficient accuracy to determine if it actually was increasing. We can today assess that the annual change then would have been less than 0.1 ppmv, which was much less than could be measured at that time. Still, his fundamental scientific work led to a much deeper understanding of key environmental processes.

Almost 100 years were to pass before Arrhenius' findings became of political interest. His discovery was a very early one and it illustrates well the fact that fundamental research often uncovers surprises that can be either destructive or beneficial. It is obvious that there was as yet no societal concern that the further development of an industrial society might lead to the impoverishment of the natural world around us. The concept of the environment as an asset beyond its provision of natural resources had not yet been recognised. Scientists, politicians and industrialists had no reason to worry about issues of this kind and the twentieth century began with an optimistic attitude towards the future.

Throughout the twentieth century, experts have been familiar with Arrhenius' work, but it was largely regarded as being something that might have to be looked at again more closely in the future. It was not until 1957 that Keeling (1958) was able to develop an accurate method of measuring the amount of carbon dioxide in the atmosphere and could show that the annual rate of increase at that time was about 0.6 ppmv and that this increase was probably due to human emissions caused by burning fossil fuels. At about the same time a renewed interest in learning about the biogeochemical cycle of carbon and climate change also emerged.

2

The natural carbon cycle and life on earth

Our knowledge about the global carbon cycle can be made more robust by making use of the condition of mass continuity, distributions of tracers and interactions with the the nutrient cycles.

2.1 Glimpses of the historical development of our knowledge

Carbon is the basic element of life. All organic compounds in nature contain carbon and the carbon dioxide in the atmosphere is the source of the carbon that plants assimilate in the process of photosynthesis. An understanding of the global carbon cycle is of basic importance in studies of human-induced climate change, not only because of the need to determine expected changes of atmospheric carbon dioxide concentrations due to human emission, but because natural changes of the carbon cycle may also have influenced the climate in the past.

The detection of the fundamental chemical and biochemical processes of relevance in this context is a most important part of the development of chemistry during the eighteenth century and the first decades of the nineteenth century. Joseph Black (1754) is credited with the discovery of carbon dioxide gas. Its real nature was, however, not very well understood until Carl W. Scheele in Sweden and Joseph Priestley in England identified 'fire air' (i.e. oxygen) a few decades later and the French chemist Lavoisier correctly interpreted the concepts of fire and combustion. When carbon burns, carbon dioxide is formed.[1]

It was not realised until well into the nineteenth century that carbon dioxide, like oxygen and nitrogen, is a permanent constituent of the air and that it is a source of carbon for plants. However, it was not then possible to measure the amount present in the atmosphere. In fact, it was not until the end of the century that the average atmospheric concentration of carbon dioxide was determined to be somewhat less than 300 ppmv. The analytical techniques were reasonably accurate, but it was not fully realised that the local carbon dioxide concentration in the air varies markedly due to its role in biological processes and also because of emissions from burning coal (From and Keeling, 1986).

When Arrhenius published his major paper on the role of carbon dioxide in the heat balance of the earth (Arrhenius, 1896a), it was not known whether or not the atmospheric concentration might be rising as a result of the increasing use of coal. Even though Arrhenius dismissed the possibility that man could influence the atmospheric concentration significantly in that way, the possibility remained in the back of the minds of several researchers during the first half of the twentieth century.[2] One may quote Lotka, who was the father of 'physical biology.' He became interested in the carbon cycle when developing this new concept. In 1924 he wrote very optimistically:

... to us, the human race in the twentieth century, this phenomenon of slow formation of fossil fuels is of altogether transcendent importance: The great industrial era is founded upon the exploitation of the fossil fuel accumulation in past geological ages ... We have every reason to be optimistic, to believe that we shall be found, ultimately, to have taken at the flood of this great tide in the affairs of men; and that we shall presently be carried on the crest of the wave into a safer harbour. There we shall view with even mind the exhaustion of the fuel that took us into port, knowing that practically imperishable resources have in the mean time been unlocked, abundantly sufficient for all our journeys to the end of time.

This he said in spite of the fact that he recognised the complexity of the issue:

But whatever may be the ultimate course of events, the present is an eminently atypical epoch. Economically we are living on our capital; biologically we are radically changing the complexion of our share in the carbon cycle by throwing into the atmosphere, from coal fires and metallurgical furnaces, ten times as much carbon dioxide as in the natural biological process of breathing. These human agencies alone would ... double the amount of carbon dioxide in the entire atmosphere ...

The first decades of the twentieth century saw the beginning of ecological thinking and in this context the circulation of carbon was also brought into focus. Vernadsky in Russia wrote his ground-breaking book on the biosphere in 1926, in which he recognised for the first time what we today call global ecology. He emphasised that '... the Earth, its atmosphere as well as its hydrosphere and landscapes, is indebted to living processes, i.e. the biota, for its present composition.'

In 1935 his colleague Kostitzin developed a quantitative model of the carbon cycle and recognised the necessity of considering in this context its interplay with the circulation of oxygen and nitrogen and in particular long-term changes in their abundance in the atmosphere and the soil. This was long before the concept of biogeochemical cycles and their interactions became a generally accepted view of the dynamics of environmental interactions. These researchers were indeed pioneers.

In England Callender (1938) addressed the question of a possible increase in atmospheric carbon dioxide due to burning of fossil fuels. He recognised that the lowest values that had been observed towards the end of the nineteenth century had usually occurred in the middle of the day and when the air was of marine or polar origin. He correctly drew the conclusion that mixing of the air horizontally as well as vertically is most efficient under these circumstances. Atmospheric concentrations were therefore likely to be least influenced by local conditions and accordingly most representative on these occasions. Callendar concluded on the basis of the measurements taken during the last decades of the nineteenth century that the most likely average concentration between 1872 and 1900 was around 290 ppmv with an uncertainty of about ±10 ppmv.[3]

This value is just slightly above what is deduced from analyses of the carbon dioxide content of air bubbles in glacier ice formed at that time. When air between the snowflakes that are deposited on the ice sheets in Antarctica and Greenland is shut off from direct contact with the atmosphere because of the accumulation of snow in the following years, air samples are created and their carbon dioxide content can be measured. By counting the number of layers that have been formed these samples can also be dated.

In the late 1950s Keeling developed a new method for measuring the amount of carbon dioxide in air and was able to show that the atmospheric concentration had risen to about 315 ppmv in the late 1950s and was increasing annually by about 0.6 ppmv (see Keeling (1960)). This is equivalent to an increase in the amount of atmospheric carbon dioxide of about 1.2 Gt C per year,[4] which corresponds to just about 0.2% of the carbon in atmospheric carbon dioxide at that time (about 670 Gt C). The annual emissions due to fossil fuel burning were, however, about 2.5 Gt. and the annual increase in the atmospheric concentration corresponded thus to merely about 50% of these emissions. The accumulated emissions due to fossil fuel burning since the industrial revolution began were then estimated to have been about 80 Gt C. These simple findings were very important and raised a number of basic questions that were addressed during the next few decades. First, there is obviously a significant exchange of carbon dioxide between the atmosphere and other natural carbon reservoirs, the sea and the terrestrial biosphere, i.e. vegetation and soils, and presumably also a net transfer from the atmosphere into these when the atmospheric concentration increases. Carbonate rocks are by far the largest reservoir of carbon on earth, but one could ask if the rates of weathering, and thus release of carbon from rocks to water and air, were small compared with the human emissions due to fossil fuel burning, and also compared with the natural flux of carbon dioxide back and forth between the atmosphere and the sea, which was of the order of 100 Gt C.

The uptake of atmospheric carbon dioxide by biospheric assimilation and the return flow to the atmosphere due to the decomposition (mineralization) of dead organic matter in the soil appeared also to be of that same magnitude. The primary interest in these matters was to determine the factors that regulate the amount of carbon dioxide in the atmosphere on the time scales of decades, centuries and millennia. One could thus then already conclude that we can largely limit ourselves to analyses of the exchange between three major carbon reservoirs, the atmosphere, the sea and the terrestrial biosphere (including soils) when investigating changes of the carbon cycle brought about by human activities.

The radioactive isotope of carbon, ^{14}C, was discovered by W. F. Libby in the 1950s, a discovery that earned him the Nobel Prize for Chemistry in 1960. A new and powerful tool for the analysis of the carbon cycle had been provided. Cosmic radiation reacts with nitrogen in the atmosphere to produce ^{14}C. The level of cosmic radiation has presumably been approximately constant over millennia and the ratio of ^{14}C to the stable isotope ^{12}C in atmosphere has also remained constant. However, when a sample of carbon is removed from the atmosphere, the amount of ^{14}C in that sample declines by about 1% in 80 years because of radioactive decay. The proportion of ^{14}C in a carbon sample can therefore be used to measure the time that has elapsed, since it was last in the atmosphere. This provides a clock that can be used to determine the rate of exchange and turnover between different parts of the carbon system. It was soon discovered that ^{14}C concentrations in the sea were significantly lower in the deeper layers, which showed that the circulation of water in the sea is a slow process (Revelle and Suess, 1957). It takes from many hundred to a few thousand years to mix the oceans.

These discoveries provided new opportunities for analysing the global carbon cycle much more stringently and not merely applying the necessary and obvious condition of mass continuity. The residence time of a carbon dioxide molecule in the atmosphere was determined to be 5–10 years (Bolin, 1960). It also became possible to analyse quantitatively the role of the oceans as a sink for the uptake of excess carbon dioxide in the atmosphere as a result of emissions from fossil fuel combustion. Fossil carbon contains no ^{14}C, since millions of years have gone by since it was buried deep in the earth's crust.

Projections of likely future atmospheric carbon dioxide concentrations could then be made under plausible assumptions about the expected rate of increase of fossil fuel use. An increase from preindustrial conditions of about 170 Gt, i.e. about 80 ppmv (i.e. 30%) by the end of the twentieth century seemed likely, as well as a possible doubling towards the end of the twenty-first century. Because of these results the possibility of a human-induced climate change could be analysed much more quantitatively during the 1970s. The work of Arrhenius was again brought into focus. Perhaps a global warming was on its way. On the other hand, observations and analyses of global temperatures at the time indicated that a slight but general cooling had occurred since about 1940. Different views of these matters were the subject of conflict for the next several decades.

The terrestrial ecosystems and particularly the forests had long been exploited and had changed drastically since the early nineteenth century as the world population increased about eightfold. Attempts to quantify these changes began in the 1970s (Bolin, 1977; Houghton *et al.*, 1983), but uncertainties were large. It was, however, quite clear that there had been an accumulated net flux from the terrestrial ecosystems to the atmosphere due to deforestation and changing land use at that time by possibly as much as 70–100 Gt C. The population increase in developing countries led to a need for more land for agriculture. In addition, the demand for wood products in developed countries and the opportunities to import wood from developing countries further increased deforestation in many developing countries. At the peak of this forest exploitation in the late 1980s, 10–15 million hectares were deforested annually, particularly in the tropical forest areas. At the end of the twentieth century the accumulated emissions were estimated to about 120 Gt C (Houghton, 1999).

However, this flux seemed to reverse in Europe and the USA during the latter part of the twentieth century because there was less demand for land due to higher yields per unit area obtained in agriculture and some reduction in the land cultivated occurred. Modern forest management also led to a build up of the amount of carbon stored in managed ecosystems. And above all the increased carbon dioxide concentration in the atmosphere stimulated photosynthesis. However, the annual emissions due to direct human intervention were still about 1.6 Gt C at the end of the twentieth century, which corresponds to about 25% of the emissions due to fossil fuel burning at that time.[5] But the sink mechanisms increased in importance and the earlier net emissions due to human interventions changed markedly to a significant net global uptake of carbon dioxide by the terrestrial ecosystems that dampened the increase in atmospheric concentrations.

2.2 A simplified view of the present carbon cycle

It is obvious that detailed knowledge about the global carbon cycle is essential in order to judge the implications of human-induced changes on the major carbon reservoirs in nature, directly or indirectly. In particular, how do such changes influence the atmospheric carbon dioxide concentration now and what may happen in the future? Detailed knowledge is also required because of the great heterogeneity of the terrestrial ecosystems. Individual nations need to understand how their biological resources would best be managed in view of their importance in the global context. The following brief overview of our knowledge at the turn of the twentieth century is given as a background for the later discussions and analyses.

Figure 2.1 shows a schematic picture of the global carbon cycle, i.e. the major carbon reservoirs in nature that we need to consider in the present context and the exchange between them.[6] Flows between the atmosphere and the oceans differ both in direction and magnitude from one part of the oceans to another. In the approximately steady state during preindustrial times these opposing flows in different parts of the world largely balanced each other. Human activities and the

The Carbon Cycle

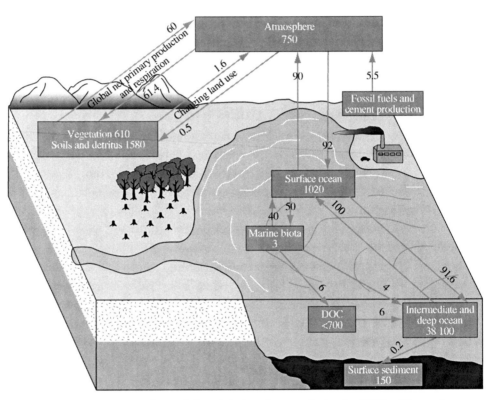

Figure 2.1   An overview of the global carbon cycle in the 1980s. The carbon content of the major reservoirs is given in Gt C (10^9 ton carbon) and the arrows show the flows between them in Gt C per year. The numbers given are rather uncertain, but the figure demonstrates the simple, overall global features of the cycle (IPCC, 1995a). By 2005 the atmospheric content had increased to about 800 Gt C, the land sink had risen from about 0.5 to about 2.5 Gt C/year and emissions due to fossil fuel burning and cement production had grown from about 5.5 to about 7.6 Gt C/year (IPCC, 2007a). (DOC = dissolved organic carbon.)

increase in the atmospheric carbon dioxide concentration have now led to a situation in which there is a net flow from the atmosphere into the oceans, the magnitude of which is equal to the difference between the gross flows that are shown. Thus the oceans presently take up part of the human emissions to the atmosphere.

The gross flows are obviously much larger than the net flows between the major reservoirs and some may wonder how this difference can be determined with the given accuracy. We cannot measure the net global flows directly with any reasonable accuracy, but based on ocean circulation models and by making use of both radioactive and stable tracers as well as measurements of the decrease in oxygen in the atmosphere when burning carbon, one can deduce that the net uptake by the oceans at present is probably 1.6 ± 0.6 Gt C per year.[7]

Even a simplified picture of the carbon transfer into the oceans must account for the rapid mixing of the surface layer and the slow transfer to deeper layers must be accounted. In fact, in the most inaccessible parts of the deep Pacific Ocean the carbon content has not yet changed at all in spite of the net flow of carbon into the ocean surface layers induced by human activities during the last few hundred years.

Two processes that act in opposite directions bring about a net vertical transfer of carbon within the ocean:

- Plankton growth takes place in the uppermost 10 m, where solar radiation is available. When plankton die, they settle slowly into the deeper layers, decomposing and dissolving on their way downward; this is *the biological pump*. It does not seem likely that the rate of photosynthesis, and therefore this transfer of dead organic matter and accordingly carbon, has been changed by human activities except in some coastal areas that have been fertilised by nutrients leached from the land. The rate of primary biological production is largely limited by the amounts of nutrients available, which have not yet been influenced significantly on a global scale by human activities. There is also a *solubility pump*, because the solubility of carbon dioxide is greater in the cold water that sinks than in less salty water.
- The downward transport of organic carbon by the biological pump and inorganic carbon by the solubility pump means that the amount of carbon in the intermediate and deep waters of the oceans is increased. On the other hand, oceans currents and turbulent motions, in particular, *transport carbon upwards* from the carbon-rich layers at greater depths.

During quasi-steady preindustrial times the global net flow due to these two different processes was close to zero. But enhanced concentrations of dissolved inorganic carbon in the surface layers due to the direct inflow of carbon dioxide across the sea surface from a carbon-enriched atmosphere leads to a decrease in the gradient of dissolved inorganic carbon between the upper and lower layers and therefore a reduction of the turbulent upward flow. The two opposite transfer processes in combination therefore now maintain a net transfer of carbon downward. The process is slow but still reduces the rate of build up of carbon dioxide in the upper layers of the ocean and thereby enhances the inflow from the atmosphere. Thus, the values that are shown in Figure 2.1 have been derived with the aid of available observations as well as from consideration of what we know about the physical, chemical and biological processes that regulate the carbon cycle. Information obtained from the simultaneous consideration of the levels of ^{14}C and radioactive hydrogen (i.e. the tritium that was formed when hydrogen bombs were tested) provides an internal 'clock' for the analysis, and better overall internal consistency of the description of the carbon cycle has been ensured in this way.

The atmospheric carbon content has increased from about 590 Gt C (about 280 ppmv) 150 years ago to about 800 Gt C (about 380 ppmv) in 2006. This increase of about 210 Gt C should be compared with the magnitude of the total emissions due to fossil fuel combustion during this time, which is now estimated to have been about 325 Gt C. In addition, there has been a net input of carbon to the atmosphere due to deforestation and changing land use, estimated to have been about 140 Gt C in 2006. Thus the air-borne fraction of the total emissions has been about 45%.

The terrestrial system has changed markedly during the last 150 years. Deforestation in both Europe and North America was the prime cause of a slow increase of atmospheric concentrations during the early parts of the nineteenth century. By the middle of the century the growing use of fossil fuels had raised the rate of increase somewhat. Use of fossil fuels gradually became the dominant cause of the increasing atmospheric concentrations of carbon, and reached a value of about 1 Gt C per year in the 1930s. Atmospheric carbon dioxide concentrations were then about 300 ppmv. After the Second World War the use of fossil fuels increased much more rapidly and gradually became a major carbon dioxide source, but the annual increase of atmospheric carbon dioxide seemed surprisingly modest in the early 1990s, even when the carbon dioxide uptake by the oceans was included. There was obviously some 'missing sink' that had not been considered adequately. However, the increased atmospheric carbon dioxide concentrations would enhance photosynthesis and changing land use and regrowth of secondary forests might be more important sinks than expected. Actually, early in the twenty-first century it became obvious that in order to balance the increasing emissions from the use of fossil fuels, the terrestrial biosphere had to serve as a very significant sink, 2–3 Gt C per year in order also to make up for the emissions due to deforestation in the tropics that amounted to at least 1.5 Gt C annually. There are primarily three factors that make the terrestrial ecosystems serve as an important carbon sink[8]:

- Regrowth of forests as a result of improved forest management in areas where substantial deforestation occurred 50–150 years ago, i.e. primarily in temperate latitudes of the Northern Hemisphere, e.g. in Europe and North America.
- The rate of photosynthesis in forests may in the past have been limited by insufficient water supply. Higher atmospheric carbon dioxide concentrations and in some areas also an enhanced nutrient supply due to industrial activities stimulate photosynthesis in plants by increasing water use efficiency.
- The warming in recent years at high latitudes in the Northern Hemisphere in combination with the higher atmospheric carbon dioxide concentrations may also have contributed to enhanced growth, particularly in Siberia and Canada.[9]

The annual emissions due to fossil fuel burning, which reached 6.3 Gt C per year in 1990, have since increased markedly to about 7.2 Gt C per year early this century and are now (2007) approaching 8.0 Gt C per year. On the basis of the observed changes in the atmospheric concentrations of carbon dioxide the annual average increase is determined to be 3.8 ± 0.2 Gt C, i.e. about 45% of the total emissions.

Because of the many different processes that play a role in the exchange of carbon between the different reservoirs (see Figure 2.1), an approximate balance can be established between the total human-induced emissions into the atmosphere (about 8.0 Gt C per year in the 1990s and above 9.0 Gt C per year in 2007) on one hand and the uptake by the terrestrial ecosystems and the oceans and the increase of atmospheric concentrations on the other.

In addition, it is important to note that the terrestrial ecosystems differ markedly from one part of the globe to another. In tropical forests most of the carbon is found in living organic matter, i.e. in the trees. Forest soils contain less carbon, but a closed and rather rapid circulation of carbon and nutrients is maintained, which, however, necessarily requires an adequate water supply. Organic matter in the soils decomposes quickly, but is replenished by dead organic matter from the forest, i.e. litter, returning carbon dioxide to the atmosphere. The tropical ecosystems are sensitive to large scale deforestation because soil deterioration may prevent the rain forest ecosystems reestablishing themselves, particularly where there is a move towards a warmer climate.

Boreal and temperate forests, on the other hand, usually grow on more carbon-rich soils and the local circulation of carbon and nutrients is slower. This is where we find extensive carbon deposits in the form of peat and permafrost land. Deforestation and the use of the land for agriculture, which occurred in Europe and North America during the nineteenth century and the early part of the twentieth century, meant a substantial loss of carbon to the atmosphere. This contributed significantly to the early build-up of carbon dioxide in the atmosphere. As previously mentioned this process seems to have been reversed during the latter part of the twentieth century.

The exploitation of the terrestrial ecosystems raises a number of questions about sustainable development. The decrease in the area covered by forests and the loss of organic matter in the soils in the course of the expansion of agricultural land has not only meant a decrease in the storage of carbon, but also an impoverishment of the land and the reduction of biodiversity in several areas. The key questions are therefore: Will the storage of carbon in terrestrial systems continue or will the storage merely be temporary and the terrestrial systems again become a carbon dioxide source and thereby increase the rate of carbon dioxide enhancement in the atmosphere? How will the gradually changing global climate influence the exchange of carbon dioxide between the atmosphere and the terrestrial systems? We do not yet have reliable answers to these questions.

We learn from this brief overview of the global carbon cycle and the ongoing changes of the terrestrial ecosystems that several major environmental issues are closely interlinked. This fact must be kept in mind when planning human activities that may change the role of the terrestrial ecosystems in the global carbon cycle. Knowledge of the dynamics of the carbon cycle is of fundamental importance in planning for agricultural land and forests to be used sustainably. Interaction is therefore needed between the scientific community and those that are currently using our natural resources unsustainably.

3

Global research initiatives in meteorology and climatology

Two decades of efforts to develop global research programs in meteorology and climatology led to the formation of the World Climate Research Programme, WCRP, in 1980.

3.1 Building scientific networks

3.1.1 The formative years

On 1 April 1960 the USA launched its first meteorological satellite, TIROS 1. It was a remarkable experience for people to be able to view the earth and its atmosphere from the outside. The bluish colour of our planet fascinated observers and a number of well-known features of the circulation of the atmosphere became visible through the cloud formations that they create. Most of what one could see was familiar to the meteorologists. It appeared so consistent with the knowledge that had been taught in basic courses for years. Nevertheless, another dimension had been added. A very effective new tool for observing the weather had become available. The weather services were soon engaged in trying to find out how this new information could best be exploited. Scientists sensed that a new era in meteorology and climatology had begun.

The event was also of profound political importance. A satellite had been launched that might be used internationally for peaceful purposes. Had a new opportunity thereby been opened in the race between the USA and the USSR to be in the lead in space? Would this indeed be the beginning of peaceful global cooperation? President J. F. Kennedy seized the opportunity. In an address to the General Assembly of the United Nations (UN) in 1961 he called on the countries of the world to exploit this new tool jointly.[1] However, the international crisis in which nuclear weapons were shipped from the USSR to Cuba soon brought a return of the very frosty relations between the USA and the USSR that had lasted since the end of Second World War in 1945.

19

This appeal to the meteorological profession was pragmatic in the sense that the weather services were put in focus, rather than the scientific community. The World Meteorological Organisation, WMO, quickly set up a small task force consisting of just two individuals, the directors of research at the weather bureaux in the USA and USSR, Harry Wexler and Victor Bugaev respectively. Their joint recommendations led to the birth of the World Weather Watch, WWW. A WMO advisory committee on research was also constituted and began a more thorough analysis of how to use this new tool operationally.

Leading US atmosphere scientists were, however, worried. They argued very strongly that the effective use of the observations from a set of orbiting satellites justified a major research effort. Although this certainly was true, it also signalled a battle for resources within the USA. Perhaps it was also an early realisation of the increasingly important role that science might play with satellites orbitting around the earth.

The previous decade had seen a remarkable development in theoretical meteorology. New methods using electronic computers for quantitative weather forecasting were being developed. These efforts had begun in the late 1940s under the leadership of Jule Charney and John von Neuman at the Institute for Advanced Studies in Princeton, NJ, with support from Carl-Gustaf Rossby, first at the University of Chicago (until 1947) and then at the University of Stockholm until his death in 1957. When the UN agreed in 1961 on a resolution to use satellites for observing the weather from space, about a dozen scientific groups were engaged in work of this kind in the USA, Europe, Japan and Australia. They asked themselves: how can a joint long-term global research effort further stimulate our ongoing work? The US National Academy of Science (NAS) formed a panel under the chairmanship of Robert White to explore what could and should be done and Charney was asked to prepare a position paper.

In the summer of 1962 the Norwegian Geophysical Society organised a conference in Bergen to celebrate the centenary of the birth of Vilhelm Bjerknes, the grand old man of the science of meteorology (who had worked with Arrhenius in Stockholm in the 1890s). On that occasion Charney led a discussion about key scientific issues that had arisen in the light of the prospect of using satellites to provide new data for weather forecasting. He returned to the USA with strong support from scientists all over the world for the engagement of the scientific community in the pursuit of this issue. Improvement of weather forecasting requires the development of a better understanding of the dynamics of the general circulation of the atmosphere, a fundamental research topic that was also very relevant in the field of climatology.

Lobbying in Washington was successful and an additional resolution was adopted by the UN General Assembly late in 1962, calling on the scientific community to contribute and supplement the initiatives that were already being organised by the WMO.[2] The responsibility for developing an expanded programme of research was given to the International Council of Scientific Unions, ICSU, and particularly to its International Union of Geodesy and Geophysics, IUGG. A balance had thus formally been struck between, on one hand, those wishing to exploit this new technology as quickly as possible to improve weather forecasting and, on the other, those emphasising strongly that enhanced scientific efforts would be required in order to make use of this new tool wisely. This left the question of how should a joint effort best be organised.

IUGG used to hold a general assembly every third year and one was scheduled for the summer of 1963 in San Francisco. I attended this meeting and gave a presentation about using tracers for studies of the ocean circulation. A special session on the use of satellites for meteorological observations was organised at the meeting. After quite extensive discussions it was decided that IUGG should launch a truly international effort in order to prepare for the use of satellite technology in studies of the general circulation of the atmosphere and to develop new methods for weather forecasting. The task of starting such an effort was given to the IUGG Bureau and I was elected a member of that Bureau with responsibility for dealing with this matter.

It was obviously important to establish proper working relationships with the WMO advisory committee. In addition, however, ICSU had already in 1958 established the Committee on Space Research, COSPAR. There was concern that there was a danger of duplication of efforts and perhaps even rivalry between a new committee, created by IUGG, and COSPAR which was already functioning. The views of the IUGG were, however, forcefully presented to the tenth general assembly of ICSU in Vienna in November 1963 and it was agreed that an interunion committee on atmospheric sciences (CAS) with IUGG as the parent organisation be formed by ICSU. Its composition was agreed upon by ICSU and IUGG jointly.

It was also agreed that COSPAR should be called upon as a partner in this undertaking. The next COSPAR meeting was scheduled for May 1964 in Florence. I asked interested scientists to meet and discuss the matter on that occasion and also invited representatives of the WMO to attend. The meeting was, however, a failure. The COSPAR scientists were largely physicists and chemists studying the upper atmosphere, i.e. the mesosphere, the ionosphere, and outer space. COSPAR Working Group VI had been created in order to provide the means for international cooperation in this work. Many of the working group were engaged in developing instruments to be used in rockets and satellites. The WMO advisory committee, on the other hand, was closely tied to the interests of the weather services, i.e. improving the means of weather forecasting. It became obvious that the IUGG should focus on the fundamental physical problems of radiation, cloud physics, turbulence and especially studies of the large scale dynamics of the troposphere and stratosphere, i.e. the general circulation of the lower atmosphere, while COSPAR's expertise should be used to analysis alternative future observational networks. This would later indeed become most important.

During a visit to the USA in the autumn of 1964 I met in Cambridge, MA, with some leading US scientists who wished the IUGG initiative to be pursued vigorously, among whom were Jule Charney and Richard Goody, Harvard University, and Thomas Malone, chairman of the Committee on Atmospheric Sciences of the US NAS. We discussed names of potential members of an IUGG committee and on my way back to Sweden I visited the president of ICSU, Professor H. W. Thompson, at Oxford University. Later that same year a formal decision was taken by the ICSU to establish a CAS and I was asked to serve as its chairman.

In retrospect, the creation of the CAS can be seen as the beginning of the development of a series of global research programmes in the field of the environmental sciences, which have been of fundamental importance for securing resources for global research efforts during the last 40 years. It was through the work of the CAS that the scientific communities in meteorology and climatology became recognised as potential users of satellites developed for peaceful purposes, just as President Kennedy had said a few years earlier. The key to success in this work has been to focus on projects that necessarily require international coordination. At that time there was not primarily a need for detailed overviews of all the scientific issues at stake, i.e. to provide comprehensive analyses, but collaboration was essential in order to define the observational requirements for testing and further developing existing models of the general circulation of the atmosphere.[3]

At the first CAS meeting, which was held in 1965 at WMO Headquarters in Geneva, its objective was defined as developing 'an entirely research oriented co-operative international meteorological and analytical programme with the goal of producing a vastly improved understanding of the general circulation of the global atmosphere ...' The programme was called the Global Atmospheric Research Programme, GARP. It was further proposed that '1972 be designated as a twelve month period for an intensive, international, observational study and analysis of the global circulation of the troposphere and lower stratosphere (below 30 kilometres).'

The WMO advisory committee also endorsed these recommendations. It is of interest to note that this advisory committee had not then proposed any specific research efforts although it had already been in existence for more than three years. Their work had obviously been focused on efforts of more immediate importance such as support of the WWW. Their support of the CAS initiative was most welcome and represented an important step towards close cooperation between ICSU and WMO.

The CAS began detailed planning of a programme at this first meeting and was greatly assisted in this work during the following two years by the analyses that had been initiated by the US NAS. The NAS' study of the feasibility of a global observation and analysis experiment, led by Charney, was at the heart of the discussions.[4] COSPAR contributed to the planning through its Working Group VI under the chairmanship of Morris Tepper (USA). The proposed schedule which called for a global experiment to start in 1972 was, however, unrealistic and GARP experiments were not ready for launch until the late 1970s. The recommendations show, however, the very proactive attitude of the members of the CAS.

The CAS felt strongly that it was essential to invite a larger group of scientists to discuss in more detail the plans for GARP that were being developed. I was asked by the Committee to organise a two-week study conference that took place near Stockholm in June–July 1967. About 70 scientists took part and for the first time a number of well-known researchers from the USSR also attended.[5] The WMO was invited to be one of the organisers of the conference, but the secretary-general of WMO, Arthur Davis, limited its involvement to just that of a cosponsor, for what reason one may wonder. To me it seemed that the ambitious plans for GARP were felt to be a threat to the WMO hegemony in coordinating global observational efforts in meteorology and indirectly a criticism of the slow pace in advancing a program as part of the WWW.

The report from the conference analysed in much detail what would be required to build global models of the atmosphere, with the aid of which its general circulation could be better understood and thereby more advanced models for weather forecasting be developed. The prime aim was obviously to define a global observing system, which would be required to achieve these goals. As chairman of the CAS I submitted to ICSU a report for careful and urgent consideration by IUGG and ICSU and transmitted it to WMO which cosponsored the study conference and necessarily had to play an important role in the future planning and organisation of GARP.

The response was quick. The matter had actually been discussed at the WMO Congress in May that same year and ICSU was similarly ready to become engaged in a cooperative effort. It was decided to launch GARP and to create a joint organising committee, JOC, whose 12 members were appointed by the two parent organisations.[6] A formal agreement was signed in Rome in October of that same year. The secretariat of the JOC was to be located in Geneva, but not in the WMO premises. The protection of the scientific integrity of the committee was a prime reason for that decision, as was stressed by Professor Garcia from Argentina, who became the executive officer of GARP.

It is also noteworthy that I had, as chairman of the committee, the power to use the available funds without formal agreement by the two parent organisations, as long as expenditure was within the overall budget as proposed by JOC and agreed by ICSU and WMO. The WMO was, however, responsible for keeping the accounts. The JOC thus became a truly scientific committee with considerable financial resources, particularly when Thomas Malone became the treasurer of ICSU and was able to secure financial support directly from some US private foundations.

When the JOC held its first meeting in Geneva in April 1968 the task as outlined by CAS was spelled out carefully.[7] The key elements were specified as

The Global Atmospheric Research Programme (GARP) is thus a programme for studying those physical processes in the troposphere and stratosphere that are essential for an understanding of:

a. The transient behaviour of the atmosphere as manifested in the large-scale fluctuations that control changes of the weather; this would lead to increasing accuracy of fore-casting over periods from one day to several weeks.
b. The factors that determine the statistical properties of the general circulation of the atmosphere, which would lead to better understanding of the physical basis of climate. This programme consists of two distinct parts, which are, however, closely inter-related:
   i. The design and testing by numerical methods of a series of theoretical models of relevant aspects of the atmosphere's behaviour to permit an increasingly precise description of the significant physical processes and their interactions;
   ii. Observational and experimental studies of the atmosphere to provide the data required for the design of such theoretical models and the testing of their validity.

The prospects for advancing the task as defined by the committee were very good. Computer capability was increasing rapidly and satellite technology was in the midst of a remarkable development. Recall that the landing of men on the moon by the USA took place just one year later. Suomi, a member of JOC, had developed the concept of how to use geo-stationary satellites for cloud observations. The proposed research was generally judged to be of direct value to society. The optimism of the post-war years was still a driving force. For the first time the issue of climate research was also spelled out explicitly as a task to be pursued by the CAS, although it would be a few years before specific actions were initiated. It is also noteworthy that the responsibility for developing this programme was given to 12 prominent scientists, all of whom were active in research. In my opening statement as convenor of the meeting in 1968 I stressed that in our common efforts to develop the GARP (JOC, 1968) '... each member of JOC was selected as an individual and not as a delegate of his country and that no member should consider himself as representing either ICSU or WMO, but rather help both organisations ...' and emphasised a dual responsibility. I stated that the committee should work out

... detailed programmes and co-ordinate the experts who will be called upon to serve as consultants or to participate in Working Groups or Study Groups, but on the other hand, it also had a responsibility 'upwards' in presenting the programmes and plans for consideration by the Executive Committees of ICSU and WMO, the ICSU General Assembly and the WMO Congress, and eventually by the United Nations, in such a way that the appeal to the world to co-operate in such a formidable scientific enterprise be at the same time clear, convincing and effective.

It is clear from the report that the scientific ambitions of the members of the committee were high. My reference to the UN was a recognition that the committee implicitly also had a responsibility to pursue the ideas that the UN General Assembly had expressed in its resolutions in 1961 and 1962. The formation of a joint committee by ICSU and WMO was a recognition of the preparatory work that both the ICSU and the WMO committees had carried out during preceding years: there was cooperation rather than competition.

3.1.2 The GARP tropical experiment begins

The time was not ripe to launch a truly global experiment immediately. Observations from tropical regions were sparse and the dynamics of tropical disturbances and their role in the exchange of heat and water vapour within the atmosphere as well as between the atmosphere and the oceans was poorly understood. Not least, there was a need to find out how the small-scale convective systems are organised and contribute to the formation of larger-scale disturbances. It was not yet realistic to aim for a dense network of observations all around the world in a global experiment as was foreseen as the ultimate project. But it was felt that sufficient knowledge and understanding might be obtained over a few months from a dense network of observations covering one or a few limited areas. This might then be useful in the interpretation of observations from a sparser network of global data.

Therefore, at an international meeting called by the WMO in 1970, the JOC proposed that a tropical subprogramme should be the first project to be implemented, while the planning for a global experiment continued. This was indeed ambitious in light of the advanced set of observations that was aimed for in the tropical experiment: these required two geo-stationary satellites, a dozen well-instrumented aircraft, two of which had to be long-range jets, and some 20 ships to establish a network of ocean stations (JOC, 1972).

After a year's delay, the experiment, which was called the GARP Atlantic Tropical Experiment (GATE) finally began in 1974. It had originally been intended to conduct the experiment in the tropical Pacific Ocean but this was vetoed by the US military authorities. In retrospect one might wonder what would have been discovered about the El Nino phenomenon in the Pacific in the 1970s, if the original plans had been realised. It would be another 15 years before the tropical Pacific was similarly well observed.

3.1.3 The GATE as a starting point for global climate studies

The prime task for the JOC was to improve the observational network in order to provide data for testing the models that were being developed for weather forecasting. This was, however, also to be an important prerequisite for the development of climate models, but it did not seem meaningful to address the climate issue in all its complexities to begin with. In addition to GATE there were also several other subprogrammes that were very important for the fulfilment of the general GARP objectives, i.e. studies of 'air–surface interaction' and 'atmospheric radiation.'

The JOC began the planning of a First Global GARP Experiment, FGGE, at the very beginning of its existence with the aim of launching an experiment in 1972. A meeting of representatives from participating countries did not, however, take place until 1970, on which occasion they were asked to support the planning so far and to commit national resources for the common purpose, but at that time the GATE necessarily, and rightly so, was given priority.

It is, however, noteworthy that by then almost 10 years had gone by since Charney had outlined in Bergen the new possibilities that computers and satellites might provide. Much had, of course, changed during these years, but the aims remained high (JOC, 1971, 1972). The plans for FGGE included the use of four geo-stationary satellites for observations of the tropics and the subtropics, and two polar orbiting satellites to achieve global coverage. In addition, the poor coverage of surface data in the southern oceans was to be improved by free-floating buoys communicating with the world data centres via satellites. Another novel observational platform was high-flying drifting balloons which provided the means to compare satellite observations and in situ measurements, with focus on the Southern Hemisphere.

However, it was seven more years before the elaborate plans were realised: the experiment took place between 1 November 1978 and 30 June 1980. This is not the place to describe these efforts in detail, but FGGE greatly advanced our knowledge.[8] In retrospect it is interesting but also sad to note that there has not been such a complete programme of atmospheric observations since then. Today, early in the twenty-first century, we are trying hard to prevent an ongoing deterioration of the surface-based observational system. Paradoxically, there is at the same time an extraordinary interest in defence against natural disasters and the threat of a human-induced climate change due to the emission of greenhouse gases is constantly increasing. This indeed signifies a remarkable change in the politics regarding scientific and technological development for the pursuit of environmental studies during the last few decades. The scientists leading the GARP efforts were lucky in that the necessity of understanding the fundamental scientific issues in order to be able to develop new methods for weather forecasting was obvious and recognised. The reduced resources in later years are, however, to a considerable degree due to the increasingly complex political issues that the world only gradually became aware of in the 1990s.

3.2 Concern for the environment reaches the political agenda

While the JOC began organising the scientific community to develop a global research programme for understanding the global circulation of the atmosphere, the global climate, its variations and other key environmental issues gradually reached the political agenda. Largely due to a Swedish initiative, the UN decided in 1968 that a UN conference on the human environment be organised in 1972 and Stockholm was chosen as the venue. Although the possibility of a human-induced change of the global climate was one justification for calling this conference, it was not the prime one. Air and water pollution at local and regional scale were much more in focus politically. Above all, the acidification of precipitation, fresh water and soil due to emission of sulphur dioxide from burning oil and coal had first been recognised in 1967 in Sweden and now gained attention at the conference.

Preparations for the UN conference of 1972 began in 1969, when Carroll Wilson at Massachusetts Institute of Technology in the USA took the initiative and organised a study of critical environmental problems (SCEP, 1970) in order to provide an up-to-date assessment of environmental issues. It was perceived by several of the participants of this conference that another, more thorough, study of man's impact on climate, SMIC, was much needed. An organising committee was formed, again with Carroll Wilson in the chair, and some 30 scientists from 14 countries participated in the workshop held in Stockholm in 1971.

The SMIC report[9] was a careful assessment of available knowledge about the global climate system and its changes. It was noted that the mean temperature of the Northern Hemisphere had increased by a few tenths of a degree centigrade during the first 40 years of the twentieth century, but declined thereafter and the question was asked if this was an expression of natural variation or if human (anthropogenic) emissions might also have been of importance. Manabe and

Wetherald's analysis a few years earlier (Manabe and Wetherald, 1967) of the role of carbon dioxide in the heat balance of the earth indicated that a doubling of the carbon dioxide concentration in the atmosphere would lead to a warming of about 2 °C i.e. just about one third of the value given earlier by Arrhenius (1896a). Their result, based on the consideration of a single vertical column of air, was, however, hardly more reliable than that of Arrhenius and there were still insufficient data to validate it. If the results by Manabe and Wetherald (1967) are accepted, the observed increase of the global mean temperature early in the century was, however, probably not the result of an increasing carbon dioxide concentration. The total increase of carbon dioxide in the atmosphere in 1940 was probably less than 10% and the associated human-induced temperature increase probably not much more than about 0.1 °C.

Some preferred to explain this warming and levelling off during the first half of the twentieth century as due to the almost complete absence of volcanic eruptions from 1912 (Katmai) until the 1960s. Volcanic dust reduces the amount of solar radiation that is transformed into heat, but it was not well known by how much. The SMIC report did not provide a basis for considering a human-induced change of the global climate as an imminent threat, although a long-term change was not excluded. This was also essentially the conclusion at the UN conference in Stockholm in 1972.

The report of the 1972 conference by Barbara Ward and René Dubos, which was written for the educated layman (Ward and Dubos, 1972), gives a simple description of the climate issue. They also emphasised the need for more efficient ways of reaching decisions on global issues. They foresaw in their concluding section on global climate change that perhaps not even the sum of all prudent decisions would be sufficient to avoid a major climate change. They felt that human global interdependence in this regard was beginning to require a new capacity for global decisions and attention and that coordinated efforts for overview and research were required.

This comment is interesting in the light of what has happened since then. The need for a well-coordinated overview of available knowledge about the environment and development was put forward in 1987 by the report of the World Commission on Environment and Development (1987). This was also the main objective when creating the IPCC, in 1988. Ward and Dubos were indeed right.

3.3 The Global Atmospheric Research Programme becomes engaged in the climate issue

At its eighth session in March 1973 the JOC (1973) responded to Recommendation (79d) from the UN Conference in Stockholm, in which the view was expressed that WMO in cooperation with ICSU should:

continue to carry out the GARP, and if necessary establish new programmes to better understand the general circulation of the atmosphere and the causes of climate change and whether the causes are natural or the result of man's activities.

The JOC formulated a set of guiding principles for such work by the committee:

The two GARP objectives are strongly related to each other. For consistency with the philosophy which has been inherent within GARP, GARP climatic studies should concentrate on those aspects which lend themselves to physical–mathematical (numerical and analytical) model studies.

The accomplishment of the second objective should not be considered as implying that it will be possible to predict climate changes. It does, however, imply that we will be able to understand the mechanisms that are responsible for the climate fluctuations and to determine the nature of the change in climate caused by given external or internal stimuli (man-made or natural) to the atmosphere–ocean–earth system.

It was further decided at the eighth JOC session that an international study conference on the physical basis of climate and climate modelling should be organised. I was given the responsibility of organising this conference, which took place near Stockholm in mid-1974. This was an opportunity to involve key scientists, who had not yet taken an active part in the work of GARP, and thereby to widen the scientific basis for dealing with the climate issue.[10]

The study conference represented a major step forward towards formulating a global research programme leading to a much deeper understanding of the general circulation of the atmosphere and the oceans and thus the climate system. The stage for planning a GARP climate programme was set. The presentations and discussions also provided an overview of our knowledge, three years after the SMIC. A few new findings deserve particular attention (see JOC (1975)).

As already mentioned, in 1967 Manabe and Wetherald had deduced the expected change of the surface temperature due to a doubling of the atmospheric carbon dioxide concentration, using a simple model of the heat balance in a vertical column through the atmosphere. Now Manabe (1975) presented the results obtained by employing for the first time a three-dimensional model (general circulation model, GCM) of the atmosphere, thus taking into consideration geographical differences and indirect effects, for example those due to changes of the distribution of snow and ice. Other assumptions were the same as those made before by Manabe and Wetherald (1967). Manabe was, however, now able to reproduce for the first time the spatial distribution of the mean temperature in winter and summer without enhanced greenhouse gas concentrations quite well, which of course is required in the validation of any climate model. While Arrhenius had estimated the warming due to doubling of the atmospheric carbon dioxide concentration to be 5–6 °C and the one-dimensional model used by Manabe and Wetharald in 1967 had yielded a warming by merely about 2 °C, the new three-dimensional model showed an average global warming of about 3 °C. The higher value as compared with the result obtained by Manabe and Wetherald was primarily due to the feedback of a reduced snow cover in a warmer world and to accounting for the distribution of land and sea on earth. It is interesting to note that Manabe and Wetherald's results for climate sensitivity still hold today, although the range of uncertainty has now been assessed: the best estimate is now 1.5–4.5 °C. This rather wide range depends on the uncertainty about a number of factors that are of relevance, in particular the interplay between the atmosphere and the oceans, reduction of snow cover in winter and also the difficulties in accounting for changes of cloud cover and cloud amount.

Lorenz (1975) gave a thoughtful and very important analysis of the concept of predictability, emphasising the difference between a sensitivity analysis as carried out by Manabe and the prediction of the gradual change of the climate as a result of increasing greenhouse gas concentrations. Charney (1975) pointed out the importance of secondary changes in the climate system, such as the reflectivity of the earth's surface and the availability of water in the soil, implicitly emphasising the need for considering the biological responses to physical changes of climate. This presentation, as well as a few others, indicated the necessity of putting physics, chemistry and biology on an equal footing in the exploration of future changes of the global climate due to human activities on earth. It was, however, quite some time before it became possible to pursue such complex studies adequately.

Although the report from this conference was important as a starting point for the GARP efforts with regard to climate change, the preparations for the FGGE were given highest priority in the work by JOC during the following few years. A second GARP climate study conference was held in 1978.[11] The aim of this conference was to assess the climate models available at the time with regard to performance and sensitivity. This was important basic work, but the likelihood of human-induced climate changes was not dealt with specifically. The conference did, however, provide a basis for the further planning of a climate research programme, which really was the key task for the JOC.

Scientists outside the community of climatologists had begun to notice the increasing amounts of carbon dioxide in the atmosphere. Ecologists and geologists entered the scene. The ICSU Scientific Committee on Problems of the Environment (SCOPE) began to bring together the available knowledge about the biogeochemical cycles, i.e. the circulation of carbon, nitrogen, phosphorus and sulphur in the environment, and their interactions. The carbon cycle, which traced the pathways of the basic element of life, was the subject of a week-long workshop in Germany in March 1977 (Bolin *et al.*, 1979). The analysis of the climate issue was broadened to give ecological and geological perspectives, which later turned out to be very important in the analyses of the carbon cycle. Although the creation of SCOPE by ICSU was an attempt to be at the service of society on environmental matters, politics did not play a part at the workshop. The outreach from the scientific community to society and politicians was still hesitant and unsatisfactory, although I did not really fully realise this myself. I remember, however, the increasing interest from the German Research Council (Deutsche Forschungsgemeinschaft) in being informed about this new development of environmental research in order to set the right priorities in the future.

A first world climate conference was organised by WMO and the United Nations Environment Program (UNEP) in 1979. The bulk of the presentations concerned the physical basis for understanding the characteristics of the climate and its changes in the past and possible human-induced changes in the future. The conference was largely technical, but an agreement was reached to appeal to the world to recognise the need for more resources for research in the field of climatology. I considered it important that the role of the biosphere was emphasised, not least in order to bring home the relevance of understanding the ongoing changes in the carbon cycle (Bolin, 1979; WMO, 1979). I was invited to give the WMO lecture at the WMO congress in 1979, which dealt with impacts of climate change on the biosphere (Bolin, 1980). It was, however, more than a decade before attempts began to consider the role of the biosphere in quantitative analyses of future changes of the global climate.

A systematic exchange of information between the scientific and political communities was thus still limited towards the end of the 1970s, except in the USA. There was little mutual understanding of the way to look at the global climate system and climate change on one hand, and the impacts on and response from society on the other. Global climate models provided the means for scientists to understand better the processes that determine the distribution of climatic zones around the world. They saw the possibility of projecting future changes induced by the varying composition of the atmosphere, but their results were still presented in terms of changes of the global mean temperature, which is a rather blunt tool for assessing the impacts of a global climate change on ecosystems and human activities. Available information did not yet stir public interest in the issue of a possible human-induced climate change.

The starting point for a politician to get to grips with the climate change issue is of course a different one. The questions are rather: Will there be more heat waves or more droughts? Will intense storms and hurricanes be more common? What about floods? And above all, will human activities on earth be threatened? The scientists were as yet giving few answers to such questions; those questions were not yet even formulated specifically.

There was, however, an increasing interest in raising public awareness. US scientists led scientific development. The US NAS initiated early assessments of what the future might bring. The climate issue had also increased the attention from ICSU and WMO, the two parent organisations of GARP in the sense that an agreement was reached in 1980 that GARP would be transformed into a committee for international cooperation in climate research. The World Climate Research Programme, WCRP, was born.

4

Early international assessments of climate change

The scientific assessments during the late 1970s and the 1980s brought the climate change issue to the attention of the UN General Assembly in 1987.

4.1 Initiation of assessments aimed at politicians and society

The first efforts to analyse climate change as a threat to humankind more specifically were made in the USA.[1] Undoubtedly, the USA was leading the development of global climate models, particularly through the work at the Geophysical Fluid Dynamics Laboratory at Princeton, NJ, and towards the end of the 1970s some interest could be observed outside the expert groups. This was partly the result of a book published by one of the participants of the SMIC conference in Stockholm 1971 (see Schneider (1976)). Similar initiatives were not common elsewhere.

There was, however, also an early interest in Sweden because of my own involvement in global environmental issues (Bolin, 1976). I was asked by the Swedish Government in 1975 to summarise available knowledge, and later that same year it was concluded in a government bill concerning future Swedish energy policy[2] that '... It is likely that climatic concerns will limit the burning of fossil fuels rather than the size of the natural resources.'

An early assessment of available knowledge regarding possible future human-induced changes of climate with the specific aim of informing a wider scientific audience was initiated by the US NAS (1977). This detailed and carefully prepared overview of the state of knowledge and recommendations for intensified research served as a basis for the US efforts for a number of years. A couple of years later the US NAS gathered together a group of eight scientists with Jule Charney as chairman to analyse the likely prospects of a human-induced change

33

of the global climate and I was invited to take part in this work.[3] This effort came about because of growing political interest in the issue from US president Jimmy Carter, based on an evaluation by his Council on Environmental Quality under the leadership of Gus Speth. The report basically agreed with the rather few model experiments that had so far been carried out with the aid of GCMs to determine the sensitivity of the climate system at equilibrium to enhanced carbon dioxide concentrations in the atmosphere. It was, however, emphasised that changes would come about gradually. The thermal inertia of the world's oceans would determine the pace of change. The ultimate warming due to emissions so far would be hidden for some time. The report concluded that

It appears that the warming will ultimately occur, and the associated regional climate changes so important to the assessment of socioeconomic consequences may well be significant, but unfortunately the latter cannot yet be adequately projected.

The Council of Environmental Quality also initiated another more penetrating analysis that was published in 1981, in which the latest scientific analyses were carefully scrutinised in collaboration with among others Dave Keeling and Robert Bacastow. Two quotations bring home very well Speth's engagement in this issue. He quotes Adelai Stevenson from 1965 saying:

We travel together, passengers in a little space ship; dependent on its vulnerable reserves of air and soil; all committed for our safety to its security and peace; preserved from annihilation only by the care, the work and, I will say, the love we give our fragile craft ... With our limited knowledge of its workings, we should not experiment with its great systems in a way that imposes unknown and potentially large risks for our future generations. In particular, we cannot presume that, in order to decide whether to proceed with the CO_2 experiment, we can accurately assess the long term costs and benefits of unprecedented changes in global climate.

The Council concluded the report with the following recommendation:

In responding to the global nature of the CO_2 problem, the United States should consider its responsibility to demonstrate a commitment to reducing the risks of inadvertent global climate modification. Because it is the largest single consumer of energy in the world, it is appropriate for the United States to exercise leadership in addressing the CO_2 problem.

However, the report had only a modest impact. Soon after its publication Ronald Reagan replaced Jimmy Carter as President of the USA and the new administration did not pay much attention to the climate change issue.

A first international assessment was initiated jointly by ICSU, UNEP and WMO in 1980 and I served as chairman of the scientific team. Scientists from a number of countries were invited to take part and the assessment was in that sense international in flavour (WMO/ICSU/UNEP, 1981).The work of the team was, however, limited to one week in Villach in Austria and the report did not really penetrate the problem in more detail than the US NAS assessment had done about a year earlier, nor was the outcome widely circulated. The 'internationalisation' of the assessment effort was not very successful.

The Climate Research Committee of the Climate Board of the US NAS initiated a more comprehensive study about a year later. Professor Smagorinsky at the Geophysical Fluid Dynamics Laboratory in Princeton led the work. A short report was published in 1982 and a more detailed one about a year later. The latter did not only analyse the likelihood of a human-induced climate change, but also addressed impacts on, for example, agriculture, water supplies and sea level, with numerous examples primarily from the USA.[4]

This report might have been the first attempt to analyse the impacts and the seriousness of the climate change issue more carefully. For example, it was concluded that US agriculture was not particularly threatened, although individual farmers in marginal areas might well suffer serious losses. On the other hand, it was also concluded that a warming of just 2 °C and a reduction in precipitation of 10% might cause considerable damage to irrigated lands that give high yields. The report also brought home the message about uncertainty in an interesting way.

There is probably some positive association between what we can predict and what we can accommodate. To predict requires some understanding, and that same understanding may help us to overcome the problem. What we have not predicted, what we may have overlooked, may be what we least understand. And when it finally forces itself on our attention, it may be harder to adapt to, precisely because it is not familiar and well understood. There may yet be surprises. Anticipating climate change is a new art. In our calm assessments we may be overlooking things that should alarm us.

After the completion of the first international assessment in 1980 I felt that it was essential that the next assessment should be more truly international and that it should go beyond an analysis of the physical aspects of climate change that had dominated the efforts so far. The assessment initiated by the US NAS, just referred to, had not yet begun. Would WMO and UNEP then support such a more penetrating analysis because of their earlier interest in the subject? In fact, on the train ride from Villach across the Alps in 1980, a group of the participants and representatives of WMO and UNEP informally discussed the possibility of doing something more substantial and I expressed the view that an analysis that was wider in scope, greater in depth and more international was most desirable.

In June 1982, Dr Mustafa Tolba, the executive director of UNEP, was in Stockholm, invited by his friend Göte Svensson, a former under-secretary in the Ministry of Environment and one of the key individuals who had organised the UN conference on the environment in Stockholm in 1972. It was midsummer day and I was invited by Dr Tolba to his hotel for discussions of a UNEP project to carry out a more extensive assessment of the climate change issue. Dr Tolba, a former professor of biology at Cairo University, was anxious that emphasis should not only be on the physical aspects of climate but that attention should be drawn to the role of the global ecosystems. Presumably the support for such an effort from within UNEP was also going to be stronger, being more in harmony with its prime tasks. It would also mean less reliance on WMO participation in this undertaking that might well have been an important aspect in the internal struggle between UN agencies.

A formal agreement between the University of Stockholm and UNEP on an assessment project was reached and the UNEP financial support was generous. It permitted a careful analysis and the engagement of a good number of prominent scientists in the field. I submitted an overall plan for the project to UNEP, which was approved early in 1983, and I was given great freedom to carry out the work that got under way towards the end of the year.

It soon became obvious, however, that it was desirable for WMO to be formally involved in the effort, as it was the UN agency for meteorology and hydrology. A modest contract was negotiated in order to achieve this. ICSU had also been left out of the agreement. I convinced them later that a most appropriate contribution from them would be to assume responsibility for the publication of the final report. This task was referred to SCOPE under ICSU with Gilbert White from the USA in the chair. It was agreed that their series of publications on environmental issues, which had run since the early 1970s, should be used as the outlet for the results. This was most important because SCOPE was well established in the scientific community and used a commercial publisher. Their books were widely circulated, more so than the regular publications by UNEP and WMO. As it turned out, the outcome of this assessment reached the international research community more effectively than the report from the US NAS published in 1983 had done. It became an important stepping stone in ICSU's attempts to broaden the WCRP and create an International Geosphere Biosphere Program, IGBP.

It was obviously important for scientists from the major countries in the world to be engaged in the assessment work. Most of the research efforts were, however, pursued in Western countries, particularly in the USA, while the input from the East and the developing countries was still very limited. To alleviate this situation somewhat early in 1985 a workshop on energy issues was organised in Tiblisi (Georgia, USSR) jointly by the Soviet Academy of Sciences and the Beijer Institute of the Swedish Academy of Sciences. Even though this was useful, it soon became obvious that the workshop was under constant surveillance by Soviet government (KGB). This was hardly surprising considering that the carbon dioxide emissions from the USSR in 1980 were well above those from the whole of Western Europe and almost as large as those from the USA and Canada. The USSR reserves of natural gas were already at that time estimated to constitute about half of the global reserves. In that sense politics entered the scene, and the meeting did not have much of an impact on the outcome of the assessment. Nevertheless, it was a valuable experience.

For a number of reasons, the 560-page report brought the issue of human-induced climate change much more to the forefront in the scientific community than earlier assessments had done, particularly amongst those engaged in analysis of the terrestrial ecosystems.[5] The following issues attracted wide attention.

A study of the role of greenhouse gases other than carbon dioxide was brought to the attention of the assessment team.[6] A closer analysis of the radiative properties of methane, nitrous oxide and chlorofluorocarbons (CFCs) showed that their collective role as greenhouse gases in the atmosphere might be almost as large as that due to the enhanced concentrations of carbon dioxide. It is remarkable that it took almost a decade for the initial evaluations of the role of CFCs by Ramanathan in 1975 to be expanded to an analysis of the implications for the future if their concentrations continued to increase at the same rate as had so far been the case. It should also be recalled that the decline of ozone in the stratosphere over the Antarctic had not yet been discovered and more stringent limitations on the use of CFCs were thus not yet being considered.

Projections of likely future concentrations of these other greenhouse gases indicated that their role might increase substantially during coming decades. The collective effect of the human-induced increases of greenhouse gas concentrations in the atmosphere might be equivalent to a doubling of atmospheric carbon dioxide concentrations before the middle of the twenty-first century. Although there were considerable uncertainties about such projections into the future, the threat of climate change became considerably more alarming. As would become clear later, however, the rates of increase of CFCs were greatly overestimated.

The likely increase in the sea level because of global warming was assessed and this showed that the expansion of seawater due to warming would probably be the most important factor during the next century. Melting mountain glaciers and the disappearance of lakes might also contribute, but melting of the Greenland and particularly the Antarctic ice sheets was not expected to add much to a rise in sea level during the next 100 years, because of their slow response even to a rather substantial increase in the temperature. The total rise was estimated to be 20–160 cm by the end of the twenty-first century, but the magnitude of the increase was very uncertain. The lower bound actually implied about the same rate of rise as had been observed during the last 100 years, but the possibility of a significantly more rapid increase could not be ignored.

It was thought that the impacts of a change of climate on ecosystems might be appreciable but lack of data precluded more precise scenarios of likely future changes. However, substantial intermediate and long-term responses in the composition, extension and location of forests were projected. A joint UNEP/ WMO/ICSU International Conference followed the completion of the report 'The assessment of the role of carbon dioxide and of other greenhouse gases in climate variations and associated impacts'.[7] The conference was opened by Dr Tolba, who gave a very powerful message about possible future disasters because of a human-induced climate change. In light of the uncertainty that still prevailed about a number of issues I was not willing to paint such a scary picture of the future. I did not agree that far-reaching actions should be taken at that stage and did not propose these in my opening statement (contrary to what was reported by Boehmer-Christiansen about 10 years later in a review of the Conference[8]). I rather wanted the conference to draw the attention of policy makers to changes that might occur and to ask them to consider carefully the need for action. I stressed that the scientific community should be careful not to become engaged in the political process beyond providing information about available knowledge.

Representatives from 29 developed and developing countries attended the conference and a series of conclusions and recommendations were agreed, but a call for immediate action to mitigate possible future changes of climate was not part of that agreement. Rather, efforts to explore the implications in more detail were recommended, not least that '... support for the analysis of policy and economic options should be increased by governments and funding agencies'.[9]

The conference also recommended that UNEP, WMO and ICSU establish an advisory group on greenhouse gases

1. to ensure that periodic assessments are undertaken of the state of scientific understanding and its practical implications,
2. to help ensure that appropriate agencies and bodies follow up the recommendations of Villach 1985,
3. to provide advice on future mechanisms and actions required at the national and international levels,
4. to encourage research in developing countries to improve energy efficiency and conservation,
5. to initiate, if deemed necessary, consideration of a global convention.

The possibility that the global climate might change as a result of human emissions of greenhouse gases was taken seriously by the conference. This undoubtedly increased the interest in the issue outside the scientific community, especially in the US Administration.

The three organisations that were engaged in the assessment agreed that a small task force be formed, the Advisory Group on Greenhouse Gases (AGGG), consisting of two scientists selected by each of the three parent organisations.[10] There was, however, also a need to establish a closer association with governments and politicians, but the means to do so were not available. The AGGG held its first meeting in July 1986 in Geneva; this meeting dealt primarily with the organisation of its future work. Most active at the meeting were Hare and Goodman. They was anxious to initiate some research projects to pursue the climate change issue more vigorously. There were, however, no discussions about ways and means to initiate broader and closer relations between the scientific community on one hand, and governments and politicians on the other.

The report of the Villach conference was presented to the ICSU general assembly in September 1986 to serve as a document for discussions about the initiation of an international geosphere–biosphere research programme. This issue had already been brought up by ICSU early in the 1980s with the aim of broadening the analyses of global environmental change and in particular to emphasise studies of chemical and biological processes as a complement to the physical aspects of global change that were in the focus of the WRCP. With this purpose in mind a symposium had been arranged in Ottawa in 1984, much to the credit of T. Malone, foreign secretary of the US NAS (see Malone and Roederer (1985)).

I had been asked by ICSU in 1985 to serve as the chairman of an ad-hoc committee to plan an international research programme focused on the global dimension of chemical and biological processes in the environment. The committee completed its report in June 1986 and it was discussed at the ICSU General Assembly in Bern, Switzerland, that same year. It was agreed that a new programme, the IGBP, be launched. Being at that time scientific advisor to the Swedish Prime Minister I was able to secure financial support from the Swedish Government to develop this international research programme, and to suggest that the secretariat of the programme be located to Sweden. This was accepted. The Royal Swedish Academy of Sciences granted the project the use of its premises. Professor Thomas Rosswall of the Agricultural College, Ultuna, Sweden, became the first executive secretary to the scientific committee that was in charge of developing the programme.

The climate change issue was not high on the IGBP agenda. Instead much effort was devoted to global ecology and biogeochemistry in general. It was important to build on the UNESCO programmes of the 1970s, Man and the Biosphere and the International Hydrological Decade, that had organised the early international research efforts. The IGBP focus on environmental issues was considerably broader than in the earlier programmes.

But an organ that provided an international meeting place for *scientists and politicians* to take responsibility for assessing the available knowledge concerning global climate change and its possible socio-economic implications was missing. In my view it was not desirable that the scientific bodies WCRP and IGBP should serve in this capacity. Their responsibility should be to address the fundamental scientific questions and plan joint global research efforts. Initiatives were taken during 1986 and 1987 to find a structure for closer future interactions between the key scientists and thereby within the scientific community at large in the field of climate change, on one hand, and politicians trying to address the global environmental issues, on the other.

The report *Our Common Future* by the WCED (1987) was instrumental in bringing the climate change issue to the attention of the UN General Assembly and finding an international structure that would make it possible to take the decisions that were needed. Communication between the assessment team responsible for the report on the climate change issue, led by the International Meteorological Institute in Stockholm, and the UN Commission had been quite successful. The informal contacts between Gordon Goodman, a member of the WMO/UNEP AGGG and the Canadian J. MacNeill, who was the Secretary of the WCED had been essential. The scientific community had brought the climate change issue to the political agenda with support from the two UN organisations UNEP and WMO. It was, however, clear to me at this stage that from now on it would be essential that the different roles of the scientific community and the political institutions were kept apart, even though close collaboration of course was essential.

Part Two

The climate change issue becomes one of global concern

5

Setting the stage

The climate change issue is brought to the attention of the UN, the Intergovernmental Panel on Climate Change, IPCC, is formed and work gets under way.

5.1 The report by the UN Commission on Environment and Development

The year was 1987. The autumn had come and the UN General Assembly had opened in New York. As described in the previous chapter the UN Commission on Environment and Development under the chairmanship of Gro Harlem Brundtland (Norway) had completed its report, *Our Common Future*, and it was about to be discussed in the General Assembly.[1]

The report painted in broad strokes a picture of a rapidly changing world and an increasing exploitation of natural resources. It referred to successes in dealing with the global issues of development:

Infant mortality is falling; human life expectancy is increasing; the proportion of the world's adults who can read and write is climbing; the proportion of children starting school is rising; and global food production increases faster than the population grows.

But at the same time it was recognised, that

... there are more hungry people in the world than ever before, and their numbers are increasing. So are the numbers who cannot read or write; the numbers without safe water or safe and sound homes, and the numbers short of wood fuel with which to cook and warm themselves. The gap between rich and poor nations is widening – not shrinking – and there is little prospect, given present trends and institutional arrangements, that the process will be reversed.

The deterioration of the environment was a major issue in the report:

There are environmental trends that threaten to radically alter the planet, that threaten the lives of many species upon it, including the human species, ... productive dry-land

43

turns into worthless desert ... More than 11 million hectares of forests are destroyed yearly, and this, over three decades would equal an area of about the size of India. Much of the forest is converted to low-grade farmland unable to support the farmers who settle it. In Europe, acid precipitation kills forests and lakes and damages the artistic and architectural heritage of nations ... The burning of fossil fuels puts into the atmosphere carbon dioxide, which is causing gradual global warming. This 'greenhouse effect' may by early next century have increased average global temperature enough to shift agricultural production areas, raise the sea level to flood coastal cities, and disrupt national economies.

As already mentioned, in the course of preparing the commission's report contacts had been established between the secretariat of the commission and Gordon Goodman (at the Beijer Institute of Human Ecology of the Swedish Academy of Sciences), who played a leading role in the earlier assessment of the climate issue by the International Meteorological Institute in Stockholm. The preliminary results of that work had thus been informally channelled into the work of the commission.

The responses to the commission's report were largely positive, even though there were objections to some conclusions and many statements in the report were vague. Those sectors of society and countries that were contributing to the changes probably did not always sense that they were implicitly criticised. The transformation of the report into an action plan appeared to be a difficult undertaking and was not dealt with in more specific terms. Preventive measures by countries were expected, but to what extent and how urgently they were needed was seldom spelled out clearly. In fact, there was not sufficient knowledge to do so at the time. The close dependence of the development of countries and their economies on the use of natural resources and particularly energy was stressed throughout the report, but also in this context little was said that could serve as a basis for action. Nevertheless, a reasonably optimistic attitude emerged:

Our report, Our Common Future, is not a prediction of ever increasing environmental decay, poverty, and hardship in an ever more polluted world among ever decreasing resources. We see instead the possibility for a new era of economic growth, one that must be based on policies that sustain and expand the environmental resource base. And we believe such growth to be absolutely essential to relieve the great poverty that is deepening in much of the developing world.

It is, however, clear that this optimistic attitude, which in itself of course was commendable, became a necessity in the course of the commission's work in order to succeed in preparing a report that would be accepted by all of its members, and in particular by developing countries. The report did not dwell much on how to deal with the controversies that might emerge between countries. Economic analyses that could shed light on these crucial issues were few. This is not surprising but is regrettable. After all, the report was the first thorough attempt to analyse the global situation as seen through the eyes of a group of people who were prominent in their respective countries and close to politics. Because the character of the problems differed from one country to another, it was difficult to agree on collective measures and changes could only come about gradually. And action would necessarily involve negotiations at a political level, which in turn would require a much more detailed factual analysis. The situation in many regards remains the same or has even worsened as we enter the twenty-first century. And I feel that a more pessimistic view of the future is now emerging, since progress in dealing with the problems as sketched has been slow or non-existent.

In the discussions of the report in the UN General Assembly the climate change issue was seized upon by some countries that feared they might be hit sooner and more severely than others. For example, the Maldives, particularly the capital Male, had recently suffered severe flooding. The possibility that the country might be an early victim of an even more devastating invasion by the sea was obviously felt to be a very real threat. The Maldives own contribution to the development of such a threat through the use of fossil fuels was obviously insignificant. The fate of the country was in the hands of the rest of the world. Botswana did not speak about flooding but rather about the increasing risk of droughts. The prolonged droughts in the Sahel region certainly also lingered in the minds of many African delegates.

The organisation of a second world conference on environment and development to take place 20 years after the first one held in Stockholm in 1972 was also on the agenda. It was decided a year later that a conference, the UN Conference of Environment and Development (UNCED), would be held in 1992 and that the venue would be Rio de Janeiro.

5.2 How to create a forum for interactions between science and politics

The AGGG, which was formed in 1986, held its first meeting in July 1986. I felt early that this small group, consisting of merely six members and called on by the three organisations, ICSU, UNEP and WMO, was not representative of the scientific community working in the many fields of interest that were involved. On the other hand, I had been impressed by the work that it had been possible to organise and the conclusions reached by the group of altogether about 40 scientists that worked for the assessment of the climate issue during the years 1983–6 (see Bolin *et al.* (1986)).

Intensified research efforts were needed, and the planning and organisation of these were being taken care of by the WCRP. It was hoped the IGBP (then in its formative stage) would take on that same task with regard to the geochemical and ecological aspects of the climate change issue. The key question remained: how should the interactions between the scientific community, stakeholders and politicians that might bring the issue forward politically be developed?

In parallel, an interesting development was under way with regard to the protection of the stratospheric ozone layer. The crucial insight that CFC gases might reduce the amount of ozone in the stratosphere had been established in the early 1970s. A temporary and voluntary agreement had then been reached in 1978 that restricted the use of these gases as the propellant gas in spray-cans. It was, however, well recognised that more controls might some day be needed, but it was not possible to proceed further at the time. In the following years UNEP arranged meetings in order to arrive at more far-reaching agreements, but progress was slow. However, in 1985 a convention was finally agreed and signed by many countries at a meeting in Vienna. It did not contain any binding commitments for the parties, but was rather supposed to serve as a framework, if more stringent measures were one day needed.

Early in 1986 researchers at the British Antarctic Survey announced the discovery of a marked decrease of the ozone layer over the Antarctic. It then took only about 18 months for a protocol with quite stringent restrictions on the use of CFCs to be agreed by the signatory countries of the convention at a meeting in Montreal, Canada. A group of scientists that had been brought together by UNEP had played a decisive role in evaluating available knowledge in the field under the leadership of its dynamic chairman, Dr Robert Watson, a scientist at NASA in the USA. The Montreal Protocol of the Convention for the Protection of the Ozone Layer became a reality. The forum for negotiations that had been established by the creation of the Vienna convention had served its purpose. Specific measures could be defined and agreed rather quickly when needed.[2] It seemed that there would be a need for a similar framework convention to address the climate change issue, even though it would not be possible to reach agreements on binding commitments for quite some time to come. The threat would have to be more specific than was yet the case and this had to be established by the scientific community. The magnitudes of the two issues were of course also very different. The economy of the industry producing the CFC gases comprised merely some few tens of billions (10^9) US dollars, while the economy of the energy industry and associated sectors of society certainly amounted to trillions (10^{12}) US dollars.

Discussions between the two UN organisations, UNEP and WMO, and some countries, notably the USA, about the development of a more effective mechanism for the assessment of the climate issue began in 1986.[3] Presumably the AGGG was not considered by these actors to have the status and composition that would be required in view of the major issues that were emerging. As already mentioned, this was indeed also my own view.

The discussions within the US administration were intense. There were great uncertainty and widely different opinions about how serious a climate change might become. Many agencies were anxious to take on a leading role, others thought it was premature, but the State Department sensed clearly that this might become a contentious issue and that politics would become involved sooner rather than later.

This naturally complicated the situation. Major societal consequences might arise and the governing council of the UNEP and executive council of WMO were apparently considered by key countries (presumably including the USA) not to be the bodies to guide jointly the assessment of such matters. The idea of another intergovernmental process emerged. Governments would then influence the work directly rather than through two organisations. A climate convention, if later agreed, would also necessarily be intergovernmental. One might wonder how much attention was given to the question: will the independent role of a scientific assessment be maintained? Not surprisingly, as it turned out later, the structure of the assessment process became very important.

The WMO congress in May 1987 and the UNEP governing council later that same year agreed that the executive heads of the two organisations should take steps to organise jointly an *intergovernmental* assessment panel on climate change. In March 1988 after further discussions and consultations the secretary general of WMO invited WMO member countries to meet and agree on the establishment of the Intergovernmental Panel on Climate Change, IPCC. An agreement had also been reached by the executive councils of the two organisations about the terms of reference that should govern the work. This paved the way for the first meeting of the IPCC in November that same year.

The key paragraphs of the WMO executive council resolution emphasised the need for a broad approach to the assessment work. Considerable emphasis was also given to the role of meteorological and hydrological services in member countries. Most of the research efforts relevant for the climate change issue were, however, pursued outside the national agencies responsible for these services. For quite some time this more limited national reference base was also reflected in the selection of national representatives to IPCC meetings. This, however, changed over the next few years and a much broader representation was achieved, not least by later including socio-economic experts.

It may also be of interest to note that in the summer of 1988 the University Corporation for Atmospheric Research (UCAR) in Boulder, CO, USA, organised its first conference on interdisciplinary research. The topic was trace gases and the biosphere, which was a central issue in the assessment of possible future changes of the global climate. The formation of an IPCC was also discussed informally and the US officials that attended the meeting asked me to express my views about the organisation of such an undertaking. My main message was that work of this kind must grow gradually and that forthcoming reports must be written by renowned scientists and in such a manner that the outcome of the analyses really would be read far outside the scientific community. The climate change issue would ultimately not concern just the scientific community, but there would be a need to reach out to the public, stake-holders, decision-makers and politicians. A clear distinction had to be maintained between a basic scientific assessment of the state of knowledge on one hand, and political negotiations to reach agreements on policies and measures, on the other.

While these discussions were going on an initiative was taken by Norway and Canada, two key countries behind the work of the WCED. After the WCED's report was submitted to the UN, Gro Harlem Brundtland of Norway, WCED's chairman, and Jim MacNeill from Canada, its Secretary General, arranged a major conference in Toronto focusing specifically on the climate change issue. Brundtland and the Prime Minister of Canada, Brian Mulroney, were the keynote speakers. Preparations for the conference had begun in the autumn of 1987, and collaboration with WMO was established. The AGGG was invited to assist. In this way I was given the opportunity to work with the organising committee. The conference took place in Toronto in June 1988. Ambitions were high, discussions lively and a number of working groups were established in order to prepare a conference statement (see Environment Canada, WMO, UNEP (1988)). A call for specific actions was agreed. Among the key actions that were enumerated were:

- Set energy policies to reduce the emissions of carbon dioxide and other trace gases.
- Reduce carbon dioxide emissions by approximately 20% of 1988 levels by the year 2005 as an initial global goal.
- Set targets for energy efficiency improvements.
- Support the work of an IPCC.

I recognised during this meeting the inadequacies of the past assessments in that specificity was lacking: the impacts of climatic change were not well understood and accordingly only described qualitatively. The projection that the global mean temperature might increase by 0.2–0.5 °C per decade, as also accepted by the conference, was based on a conclusion from the WMO/UNEP/ICSU assessment two years earlier, but had not yet been generally accepted by the scientific community. I felt that the appeal to countries to cut their emissions by 20% by 2005 was an unrealistic ad-hoc recommendation that was agreed with little consideration of the importance of energy in society and the inherent inertia of governmental institutions as well as private corporations, and that there was inadequate consideration of the needs of developing countries and the socio-economic implications of rapid transformations of the energy supply systems in the world. The need for another, more trustworthy, assessment was very obvious.

The summer of 1988 was hot in the USA. A major crop failure seemed increasingly plausible. Public discussions about a possible future human-induced change of the global climate emerged for the first time. The Committee on Energy and Natural Resources of the US Senate called for a hearing in June. A statement by James Hansen of the NASA Goddard Institute for Space Studies on that occasion caught the attention of the media.[4] He claimed that

... (2) the global warming is now sufficiently large that we can ascribe with a high degree of confidence a cause and effect relationship to the greenhouse effect, and (3) in our computer climate simulations the greenhouse effect is now already large enough to begin to affect the probability of the occurrence of extreme events such as summer heat waves ...

An intense debate amongst scientists followed and most of them disagreed strongly with Hansen's statement. The data showing the global increase of temperature had not been scrutinised well enough and there was insufficient evidence that extreme events had become more common. This was to me a clear warning of how chaotic a debate between scientists and the public might become, if a much more stringent approach to the assessment of available knowledge was not instituted.

5.3 The IPCC is formed and a first assessment is begun

Only 28 countries responded to the call for the meeting in Geneva in November 1988 in order to form a panel on climate change (IPCC, 1988).[5] Only 11 of these were developing countries, but Brazil, China, Mexico, India and Nigeria attended, as well as key industrialised countries. The climate issue was still not high on the political agenda.

The executive director of UNEP, Dr Tolba, had taken charge of the preparations for the meeting. I was asked by him to serve as chairman of the IPCC. This was a tempting opportunity, but I had not been approached before the meeting and was not prepared for the invitation. Anyhow, after consultations with the Swedish minister for energy with responsibility also for environmental matters, Birgitta Dahl, I responded positively. A vice-chairman and a rapporteur were also elected, A. Al-Gain from Saudi Arabia and J. A. Adejokun from Nigeria, respectively.

Dr Tolba proposed further that three working groups be formed, which was agreed. Their tasks were defined and chairmen elected. Much of this had been prepared beforehand and was accepted without much debate. The following structure emerged:[6]

> *Working Group* I: Assessment of available scientific information on climate change. Chairman: Sir John Houghton, head of the British Meteorological Office, UK.
>
> *Working Group* II: Assessment of environmental and socio-economic impacts of climate change. Chairman: Dr Yuri Izrael, head of the Hydro-Meteorological Service, USSR.
>
> *Working Group* III: The formulation of response strategies. Dr Frederick Bernthal, assistant secretary of state, US Department of State.

The WMO took responsibility for the secretariat and assigned N. Sundararaman to be in charge.

The choice of chairmen of the IPCC and its working groups reflected how both scientific competence and political considerations played a role. Most of the work would necessarily have to be carried out in the working groups and invitations would have to be extended to key scientists to take the lead in the assessments of present knowledge in relevant fields of research. It was also important that the working group structure was agreed. Two vice-chairmen were elected for each of Working Groups I and II, but the interest in being a member of the Working Group III resulted in five vice-chairmen being chosen. In view of the fact that only 28 countries attended the conference it was further agreed that each country could chose to be a member of one of the working groups, a precautionary action that prevented conflicts. It was striking that the meeting went very smoothly and few divergent views were expressed.

A work programme was also discussed. Malta pushed very hard for the production of a report by the IPCC in time for the UN General Assembly in 1990. There was quite some resistance to committing the IPCC to such a tough schedule, but it was finally agreed. In parallel, Malta intervened at the ongoing UN General Assembly in New York, proposing a resolution on 'Conservation of Climate as Part of the Common Heritage of Mankind' and inviting the IPCC to submit its first review on the issue of a human-induced climate change by 1990. This was agreed by the Assembly.[7] This impetus turned out to be very important. It was decided that the working groups would meet at the beginning of 1989. Work got under way very quickly. It also meant that the IPCC was recognised early by the General Assembly of the UN, i.e. by the nations of the world.

It was clear to the leaders of the IPCC that we had to develop our own procedure for how to achieve the task that had been given us. During the first couple of years we formally followed the WMO procedures when in doubt. This lack of more precise rules of procedure for a task that was going to be rather different from the ordinary WMO activities gave the IPCC great flexibility in handling matters and could be exploited to the advantage of the assessment process, but care had to be exercised. It gradually became apparent, however, that we had to become more strict and professional in our work, but this had to be achieved without losing the scientific atmosphere and integrity that was essential to be able to attract the very best scientists into the work.[8]

Few, however, fully realised at that time the far-reaching implications of this UN resolution and the challenging task that lay ahead for the IPCC. I did not foresee the leadership requirement of the position of chairman of the IPCC.

The key points in the decision by the WMO executive council read as follows (the corresponding resolution by the UNEP Governing Council is almost identical):

BEING AWARE OF:
1. the results of recent international meetings that produced an updated assessment of possible climate change and suggested actions towards developing policies for responding to climate change;
2. a need to
   (a) maintain and develop further an efficient long-term monitoring system, making it possible to diagnose accurately the current state of the climate system, the trends, and the factors having an influence on climate,
   (b) improve our knowledge of the sources and sinks of the major radiatively important trace gases ('greenhouse gases'), and develop more reliable methods for predicting their future atmospheric concentrations,
   (c) promote research aimed at closing the gaps in our ability to understand and predict the climate system, including reliable projections of the regional distribution of the expected climate change.

CONSIDERING:
1. that there is a growing international concern about the possible socio-economic consequences of the increasing concentrations of radiatively active trace substances (greenhouse gases and particulates);
2. that several nations have undertaken scientific assessments of this issue during the last few years;
3. the current and the potential involvement of national meteorological and hydrological agencies in comprehensive integrated national and regional studies of the consequences of a climatic change on natural and human ecosystems taking into account sociological and economic factors, and in developing strategies for adjustment to a climate change, especially regarding agriculture and water resources.

CONSIDERING FURTHER:
that there is an urgent need to evaluate to what extent a climate change can be delayed by appropriate national/international actions,

AGREES:
that an IPCC should be established.

It was also recommended that countries should 'take into consideration the importance of such representation being at as high a level as possible and include persons knowledgeable of science, environment and related policy issues.'

6

The scientific basis for a climate convention

IPCC works towards presenting a comprehensive analysis of a possible human-induced climate change; politicians try to understand and position themselves in preparation for likely future negotiations of a climate change convention.

6.1 Work begins

The IPCC had been formed and its task for the next 18 months had been agreed. Time was short for the completion of a full assessment. The work had to be done in a new setting and involve many people who had not previously been engaged in such activities. Measures had to be taken to structure the work and to ensure broad participation by key scientists. The three working groups scheduled meetings for the early months of 1989. Outlines of their reports were agreed, lead authors for the individual chapters selected and workshops on key topics scheduled.

The fact that the issue of climate change had been brought to the attention of the UN General Assembly implicitly meant recognition of the importance of the climate change issue and also of the IPCC itself. This became obvious at the first meeting of Working Group III in Washington in January 1989 (see IPCC (1989a)). The newly appointed US Secretary of State, James Baker, gave his very first public speech in his new position at the opening of this meeting. The majority of those attending were not scientists. There were five members of US Congress and some 25 delegates from government ministries and agencies. This was not the composition of the audience that I had imagined. It is interesting to recall a few key issues that were brought forward on this occasion.

Secretary Baker applauded the recognition of environmental problems as a transnational issue. He declared that the time had come for political action, stating that the problem was not merely a scientific one, but rather a diplomatic one involving when and where to take action. He concluded by stating that progress generally results when common interests are joined to a common understanding, and declared that the present meeting would play a crucial role in generating that common understanding.

I welcomed the involvement of the US Secretary of State but it was obvious that Baker did not realise the scope of an oncoming climate change, the scale of the efforts that might be needed, and that it was premature to rush into an action programme. I therefore stressed that it was necessary to clarify better than had so far been possible to what extent a possible future of climate change might threaten the countries of the world, both developed and developing. This will require well-directed efforts by the scientific community for a good number of years, but I finished by emphasising that an analysis of the overall policy problems would not be delayed pending the more detailed analysis of the scientific problems involved.

The US State Department had obviously taken a leading role in the development of the US strategy for handling the climate change issue and indicated its wish to continue to do so in the future. The attitude of the US Administration at that time was strikingly different to what it would be in the early years of the twenty-first century.

Just a few months later France, Norway and the Netherlands invited nations to a meeting in The Hague to address the climate issue. The prime ministers of the three countries and several other heads of state attended. There were also numerous other political gatherings that addressed the issue during 1989. No substantial progress was made, however, since the factual basis that was readily available for these somewhat impromptu arrangements was just the WMO/UNEP/ICSU assessment that had been completed in 1986. There was apparently a need and a wish amongst politicians to acquaint themselves with the climate change issue and to declare their willingness to address it, but few realised yet its full scope. This was a challenging atmosphere for the IPCC to work in, when starting to update the earlier assessments.

Initiatives were also taken to widen the awareness of the climate change issue in developing countries. The TATA Energy Research Institute, TERI, in India, in association with the Woods Hole Research Center, Woods Hole, MA, USA, invited scientists and politicians from all over the world to a conference in February 1989 in Delhi, with the aim of analysing the issue of global warming from the perspective of developing countries.[1] The participants were primarily scientists and representatives from some government agencies in Asia and the USA. The WMO/UNEP/ICSU report of 1986 again served as the scientific basis for the discussions, but few of the key scientists that were present had been involved in any of the assessments so far available. This became very obvious in the way the final report from the conference was worded. The opening sentences read as follows:

Global warming is the greatest crisis ever faced collectively by humankind, unlike other earlier crises, it is global in nature, threatens the very survival of civilisation, and promises to throw up only losers over the entire international socio-economic fabric. The reason for such a potential apocalyptic scenario is simple: climate changes of geological proportions are occurring over time-spans as short as a single human lifetime.

The report was a desperate cry for political action without an appropriate scientific analysis. Many of the 'scientific conclusions' were simply not based on reliable scientific analyses. It was also markedly determined by the approach to the issue as developed at the Woods Hole Research Center under the leadership of George Woodwell and was also influenced by the feeling in developing countries of being marginalised when issues of great concern for their development were raised. It showed very clearly the need for serious international assessments with concern for not only available scientific knowledge about the climate system, but also a well-balanced approach when analysing likely impacts of a global change and socio-economic issues.

The IPCC met again in a plenary session in Nairobi in mid-1989 to take stock of how the work had advanced so far.[2] The work plans as outlined by the three working groups were approved. Retroactively! In fact the work was already well under way. This had informally been agreed between the chairmen of the working groups and me. Two other important decisions were also taken on this occasion.

- It was obviously important to help developing countries to participate effectively in the IPCC activities. On the basis of a proposal from a subgroup with members from Saudi Arabia, Senegal, Brazil and Zimbabwe (vice-chairmen of the three working groups) the IPCC agreed that a special committee on the participation of developing countries be established under the chairmanship of Jean Ripert (France), former Deputy Secretary General of the UN. This early initiative was very important because it created a body for developing countries under the able chairmanship of a former high-ranking diplomat in the UN to coordinate initiatives and protect their interests. Ripert's respect for the scientific process was also essential in the formulations of the recommendations that were put forward by the committee in the next few years. It was recognised that, in the short term, funds for travel support were urgently needed to make it possible for delegates from developing countries to participate in IPCC activities. An IPCC Trust Fund had been established at the very beginning and the IPCC now agreed that until the completion of the first assessment a sum of one million US dollars should be set aside to assist developing countries. I judged that this cooperative spirit was essential to secure acceptance by developing countries of the work being done. Their participation was in this way very well recognised at the IPCC meetings. This undoubtedly also served as an educational process of quite some significance. The influence of the special interest of oil-producing countries was not yet apparent, but it is noteworthy that the IPCC vice-chairman from Saudi Arabia followed the development attentively.

- Common scenarios for possible future emissions were needed to determine the expected changes in atmospheric greenhouse gas concentrations. Working Group III proposed that four such scenarios be developed and adopted. At that time few attempts to construct such scenarios had been described in the scientific literature and views were divided about their relevance and applicability. The proposal was, however, agreed with the understanding that the scenarios would be annotated carefully to avoid misinterpretation and misuse and that their implications should be explained. It was emphasised that the scenarios were not predictions of the future, nor were they necessarily descriptions of desirable goals. They were simply to be considered as statements about 'what might happen' under alternative assumptions about the future in order to permit an evaluation of environmental and socio-economic consequences in an internally consistent manner.

The assessment work had now to be completed in 12 months in order to finalise the report in time for the UN General Assembly that would begin in September 1990. The IPCC planned about 50 meetings and workshops during this period of time and expected many hundreds of scientists to participate. Work progressed at a remarkable pace and in a very positive and participatory atmosphere.

6.2 Politicians are anxious to show their concern for the environment

In September 1988 the British Prime Minister, Margaret Thatcher, in a talk at the Royal Society, emphasised that the issue of human-induced climate change, as a threat to mankind, would have to be taken seriously:[3]

We are told that a warming effect of 1 °C per decade would greatly exceed the capacity of our natural habitat to cope. Such warming could cause accelerated melting of glacier ice and a consequent increase of the sea level of several feet over the next century ... It is noteworthy that the five warmest years in a century of records have all been in the 1980s – though we may not have seen much evidence in Britain.

Her talk was a general call on science to be a driving force for the further development of society and in that context she expressed a sincere wish to consider carefully the increasing threat to the environment that the process of industrialisation might become. Her intervention and engagement was interpreted as an expression of her considerable understanding and knowledge of science and technology and their roles in society, but she was seriously misinformed when referring to a warming effect of 1 °C per decade. Her motives were, however, presumably also political. It was indeed a political advantage to have a forceful environmental justification for closing coalmines and instead developing the British oil and gas resources. This would reduce carbon dioxide emissions. Producing one unit of energy by burning gas, oil and coal gives rise to carbon dioxide emissions to the atmosphere in the proportion 100 : 150 : 180 respectively.

A ministerial conference on atmospheric pollution and climate change was held in Noordwijk in the Netherlands, in November 1989 in the midst of the hectic activities in which the IPCC was involved. Altogether 67 countries were represented, most of them at ministerial level, a remarkable increase of interest since the first IPCC session just one year before. When presenting the IPCC and its work toward a first assessment report (FAR) at the conference, I remarked:

It is of course impossible to present here a more detailed picture of the challenges that are confronting us ... My contribution in the present context is primarily aimed at looking beyond the next 10–20 years that have become the prime concern in the negotiations for a conference declaration ... However important the present conference is, and however successful the outcome will be, we have not more than begun a long-term development that will markedly change our view of the interplay between humans and our global environment. The societal implications cannot yet be foreseen very well, but in the co-operation between the scientific community and politicians it is necessary to achieve as optimal a development [of the politics] as possible.

The outcome of the conference was another declaration that opened with a very well-worded summary of what we are facing. In fact most of the crucial conclusions that were later substantiated in the IPCC FAR were brought into focus at the conference and implicitly accepted by the participating countries, though no action was agreed.[4] Nevertheless, public awareness and political engagement was enhanced.

The negotiations about the formulation of the declarations from the conference were in a sense forerunners of the items that would be discussed in the organised negotiations about a framework convention by the Intergovernmental Negotiating Committee (INC) that were begun a little more than a year later. For example in paragraph (8) we recognise the elements of Article 2 in the convention that was adopted in Rio de Janeiro in 1992. It reads:

For the long term safe-guarding of our planet and maintaining its ecological balance, joint efforts and action should aim at limiting and reducing emissions and increasing sinks for greenhouse gases to a level consistent with the natural capacity of the planet. Such a level should be reached within a time frame sufficient to allow ecosystems to adapt naturally to climate change, to ensure that food production is not threatened and permit economic activity to develop in a sustainable and environmentally sound manner. Stabilising the atmospheric concentrations of greenhouse gases is an imperative goal.

The third IPCC session was held in February 1990 in Washington, primarily with the aim of reporting on progress made by the three working groups and coordinating their activities where needed.[5] This session was opened by the President of the USA, George Bush. He gave a strong message about his willingness to involve the USA in the efforts necessary to protect the environment and referred to his State of the Union address a week earlier, in which he had spoken about stewardship:

It is both an honour and a pleasure to be the first American President to speak to this organization, as its work takes shape ... A sound environment is the basis for the continuity and quality of human life and enterprise. Clearly, strong economies allow nations to fulfil the obligations and environmental stewardship. Where there is economic strength, such protection is possible. But where there is poverty, the competition for resources gets much harder. Stewardship suffers.

This crucial paragraph signals the influence of strong economical players within the USA that later were going to have an increasing influence on the US policy on global climate change. Still, Bush expressed a very positive attitude towards the IPCC on this occasion.

The United States is strongly committed to the IPCC process of international co-operation on global climate change ... Last week in my State of the Union address, I spoke of stewardship because I believe that it is something that we owe ourselves, our children and their children. So we are renewing the ethic of stewardship in our domestic programs; in our work to forge international agreements; in our assistance to developing and supporting East bloc nations; and here, by chairing the Response Strategies Working Group.

He stressed the goal '... not to work in conflict, but with our industrial sectors, to move beyond the practice of command, control and compliance toward a new kind of environmental co-operation ... well-informed free markets yield the most creative solutions ... Last December in my meeting with President Gorbachev, I proposed that the United States offers a venue for the first negotiating session for a framework convention, once the IPCC completes its work. I reiterate the invitation here ...'.

The work of the session had undoubtedly started in a positive atmosphere; the USA was going to take the issue of global climate change seriously. One is of course also struck by the very different attitude on that occasion compared with the policy adopted by President George W. Bush 12 years later.

In my address to the conference I attempted to summarise the findings of the assessment process so far, but of course with the reservation that the analyses had not yet been completed:

The large number of meetings ... as well as the President of the US greeting us at this time shows that there is now a general awareness among nations about the threat of a likely climate change ... We are now entering a new phase ... [that] ... can most easily be characterised by a few questions, that all of you, I am sure, naturally have in your mind: How will *my country* be hit? What should *my contribution* be to prevent the changes of the climate that have presumably begun but cannot be described accurately at present? How should *I act* in the forthcoming negotiations? In view of the complexities of the issue, it is obvious that these questions cannot be easily answered and there is a danger that serious negotiation . . . will only begin slowly, . . . [but] . . . there is an implicit belief in the way the request to IPCC from the ministers in Noordwijk has been worded, that adequate answers can be given soon . . . However, we have rather a lengthy process ahead of us . . . no delays in coming to grips with the question of how to deal with this issue are therefore acceptable.

As is apparent now more than 15 years later, my reflections in 1990 were well justified. Progress would not be that easy and straightforward, national interests were much stronger than the willingness to become engaged in the global aspects of the issue. I stressed a set of findings about the climate change issue that were crucial facts that should be carefully considered: good knowledge about the ongoing increases in concentrations of key greenhouse gases was available and methods for comparing their roles were well in hand but progress in the field of climate change projection had been achieved only slowly. I continued

. . . it seems plausible that the present range of uncertainty, i.e. a global warming by 1.5–4.5 °C for a doubling of carbon dioxide in the atmosphere, will narrow somewhat during the next five years, . . . but reliable regional projections require better understanding of the reasons for past observed regional changes . . . We do not even know to what extent such regional anomalies actually are predictable . . .

In the midst of the hectic activities that the IPCC was involved in my selection of this set of key findings was based on an evaluation of what might be the most important conclusions of the ongoing assessment as seen from a political point of view. In that sense politics was mixed with scientific facts. It should be stressed, however, that I left politicians and others to judge how serious a change of climate might be, the urgency of specific measures, which regions of the world and which countries might be most vulnerable, how burdens should be shared amongst countries, and how soon and at what level stabilisation of greenhouse gas concentrations should be achieved. These are primarily political issues and scientists can only provide answers to the technical and economic issues that arise.

Shortly thereafter President Bush invited representatives of the 20 most influential countries in the world to a White House conference on science and economics research related to global change (17–18 April, 1990, in Washington). Even though the FAR would soon be completed and was intended to serve as the basis for negotiating a climate convention, no invitation to attend the conference was extended to the IPCC. I was surprised and sought an explanation through my contact in the USA (Dr Robert Corell) and I was soon thereafter invited to attend. For the first time I sensed that the IPCC messages might be disturbing the formulation of a US policy about these matters.

It was clear from the organisation of the conference that it was aimed at emphasising the importance of market economy in dealing with global environmental issues and demonstrating the US leadership in this context. It could be seen as an early response to the forthcoming FAR and a preparation for the negotiations about a climate convention that would follow and to which the USA had extended an invitation to the first meeting to be held in Washington. Key members of the US cabinet presented their views and proposed initiatives to be taken. Limited time was given for discussions. There were for those reasons considerable misgivings amongst invited representatives from other countries. The USA obviously wanted to present its approach to the climate change issue without interference. Towards the end of the conference I was given the opportunity of addressing the participants at the closing dinner in the US State Department. The IPCC Working Group III (which was responsible for dealing with economic issues and was chaired by the USA) was in the midst of finalising its report. This report was, however, going to deal with the economic issues only rather briefly, primarily because of the limited number of analyses that were at that time available in the scientific literature. I emphasised, based on the discussions so far within the IPCC, a few of the key findings of the third IPCC plenary meeting in Washington. In particular I emphasised the uncertainties that still prevail and continued:

It is, however, easy to hide behind the concept of uncertainty. The more important it is to extract what we can say with some certainty: ... the inertia of the climate system is considerable, due to the large heat capacity of the oceans ... We are [therefore] committed to a further change [of the climate] of perhaps [an additional] 50%, when a change has occurred ... it will then be with us for one, perhaps several, centuries. From a practical [political] point of view a climate change due to increasing greenhouse gas concentrations is irreversible. [Therefore] ... we must not merely wait for more specific and detailed analyses. It is important to prepare for the likely need for future actions and accordingly explore the implications of limiting the future concentrations of greenhouse gases in the atmosphere to some specific concentration, although we do not yet know well how serious the impacts of the associated changes of climate might be.

Energy policy has so far always been viewed as the policy required securing the 'necessary' energy supply ... A continued exclusive consideration of the energy supply side of the problem while simultaneously trying to avoid climate change necessarily means major investments in non-fossil fuel supply systems, i.e. nuclear or renewable energy ... the characteristic introduction time for new supply systems is of the order of several decades and there are at present no economic incentives to build up such renewable energy systems. Plenty of fossil fuel is available at comparatively low costs ... We should [therefore also] consider more closely the 'energy end-use efficiency' side of the problem.

It is of course most important that market forces would be able to play a role in optimising the development towards an efficient and non-polluting society. Let us, however, keep in mind the well-known essay by Professor Hardin [published in the nineteenth century]: *The tragedy of the commons*. If the individual actor can optimise his benefits by forcing all competitors to share his costs (i.e. pay for pollution) without he himself taking the necessary precautions (i.e. reducing pollution) the economy becomes unstable and the development leads to catastrophe . . . We must also recognise the vast differences in economic strength between the players on the international scene, i.e. the nations and the large multinational companies in the world . . . There are, however, a number of political issues that arise, particularly those related to developing countries . . . Major efforts to increase economic assistance and technology transfer will be required . . . But above all negotiations about a climate convention must proceed in parallel with a continued research effort.

It was clear that only confrontation with the political factors would lead to the scientific assessments being focused on scientific issues of most political relevance. Planning action familiarises politicians with the key issues at stake. The display of the different views of participating nations was perhaps the most important outcome of the conference.

The White House conference was followed by some domestic activities in the USA. The director of the Office of Science and Technological Policy, A. Bromley, outlined the US strategy for dealing with the climate change issues.[6] The IPCC assessment served as the basis of the strategy, but he emphasised very much the issue of uncertainty, which led him to the conclusion that '. . . [the uncertainties] do make it exceedingly difficult to impose policies that may have large additional costs for specific sectors of society or for specific countries, because the affected sectors or countries can legitimately point to the uncertainties in arguing against the policies. Rather, fostering more efficient use of energy, particularly by the U.S. automotive fleet and . . . planting of a billion trees per year for five years across America [and some other largely voluntary actions], could hold the US greenhouse gas emissions at the 1987 levels until at least the year 2000.' It is obvious today, some years into the twenty-first century, that little of this was ever implemented and the US emissions of carbon dioxide have risen by more than 20% during the last 15 years.

6.3 The IPCC works towards the completion of the First Assessment Report

6.3.1 Working group I. The scientific assessment

The three working groups met to approve their respective reports in May–June 1990. Work had progressed differently in the groups. Working Group I had in a sense the easiest task in that several assessments of the science of climate change had been carried out before. One was treading on familiar ground.[7] Nevertheless, agreements were not always easy to reach. I had repeatedly pointed out to the working groups that the goal was not necessarily always to reach an agreement, but rather to point out different views when necessary and to clarify the reasons for disagreements when possible, but this was still seldom tried. However, this does not mean that uncertainties were ignored. They often appeared in terms of ranges of uncertainty; however, the reasons for these were not always explained adequately and were sometimes of course simply not known.

For example, the analyses about the sensitivity of the climate system to changes of greenhouse gas concentrations (i.e. the expected change of the global mean temperature as a result of doubling of carbon dioxide in the atmosphere) led to long discussions. Some of those engaged in modelling proposed that the uncertainty range might possibly be reduced to 2–4 °C rather than the 1.5–4.5 °C commonly given at the time. This was, however, not accepted. There were as yet no decisive modelling results that would justify such a reduction. About 20 research groups had carried out sensitivity calculations and values ranged between 1.9 and 5.2 °C. The choice of a bound as low as 1.5 °C was based on the observation that the global mean temperature had so far increased by only about 0.5 °C since early in the twentieth century, in spite of the fact that the rise in the key greenhouse gases so far was estimated to be equivalent to about a 45% increase of carbon dioxide which would imply an increase of the global mean temperature by 0.7–2.1 °C. This apparent discrepancy could not be explained by just the inertia of the climate system. The likely reduction of the temperature due to human-induced increases of aerosols in the atmosphere had, however, not yet been recognised adequately.[8] Actually this wider range of the uncertainty still remains when we now (2007) are well into the twenty-first century.

Working Group I adopted the principle of ordering their conclusions in terms of uncertainty. The executive summary accordingly had the following format, which found wide acceptance.

- We are certain of the following . . .
- We calculate with certainty that . . .
- Based on current models we predict . . .
- There are many uncertainties in our predictions particularly with regard to the timing, magnitude and regional patterns of climate change, due to our incomplete understanding . . .
- Our judgement is that . . .
- To improve our predictive capability, we need . . .
- You can never exclude surprises.

Working Group I estimated the radiative forcing in 1990 caused by increased concentrations of carbon dioxide, methane, nitrous oxide and CFC gases to be 2.5 W m^{-2}. The contributions by the CFC gases were overestimated, but on the other hand the role of ozone could not then be estimated adequately. Now (2007) it has been fully recognised that human-induced aerosols also play an important role and might reduce the greenhouse gas forcing globally by as much as 40%. This means that the projected increase of anthropogenic forcing during the twenty-first century was significantly overestimated at that time. The uncertainties of the projection of future changes of climate were, however, emphasised and the changes during the following 15 years have been within the range of uncertainty as given then.

It so happened that the summer of 1988 had been warm and dry in North America and some (particularly James Hansen at NASA) took this as a sign that human-induced climate change might already be being observed. As pointed out in the previous chapter, a firm conclusion of this kind could not be drawn on the basis of the information available at that time. In IPCC (1990b) it was rather concluded that

The size of the warming over the last century is broadly consistent with the predictions of climate models, but is also of the same magnitude as natural climate variability . . . The observed increase could be largely due to natural variability; alternatively this variability and other man-made factors could have offset a still larger man-made greenhouse warming. The unequivocal detection of the enhanced greenhouse effect from observations is not likely for a decade or more, when the commitment to future climate change will be considerably larger than today.

I note in retrospect that there was almost unanimous agreement amongst scientists at the time of the Third Assessment Report (TAR) in 2001 that a human-induced climate change was being observed.

The report of Working Group I made headlines in the media and the response of the scientific community was largely positive. There were some critical remarks from those who represented activities in society that might be hit negatively by reducing the rate of fossil fuels use. This was about the time when groups of anti-climate-change lobbyists appeared in the USA as an organised opposition to the IPCC efforts. But the IPCC conclusions were carefully worded and did not say that a human-induced climate change was under way and the issue did not yet seem very threatening to most people.

6.3.2 Working Group II. Impact assessment

The other two working groups encountered greater difficulties. Since few scenarios of climate change were available to Working Group II when pursuing its task, the impact analyses were largely qualitative and expressed in terms of sensitivities of ecosystems, sea level, etc. to prescribed changes of climate. Little could be said about geographical differences except that the expected temperature rise might on average be larger over the continents and at high latitudes, and that small island states in the tropics, particularly those situated on coral reefs, would be vulnerable to increases in sea level. This qualitative nature of the report seemed appropriate.

The finalisation of the Working Group II report[9] also led to controversies with Working Group I, because the projection of the future climate change by some Soviet researchers, notably M. Budyko, on some occasions in collaboration with Yuri Izrael. (chairman of Working Group II),[10] was not accepted by most scientists from Europe and USA because of the use of analogues to the present ongoing changes. Budyko derived his results from analyses of temperature patterns as deduced for earlier interglacial periods, which were warmer than the prevailing one. The position of Izrael on this matter caused a rather serious disagreement between him and the chairman of Working Group I, John Houghton, who raised strong objections[11]:

I was disturbed to hear that . . . the WG I report had been criticized as not being written by the best scientists in the world and that the meeting was asked to express its dissatisfaction with the WG I report. I must point out that such comments are completely incorrect and unjustified . . .

He further emphasised that it actually was written '. . . by thirty of the world's best climate scientists.'

Budyko was invited to a special meeting arranged by Working Group I for more penetrating discussions and it was finally concluded that[9]:

for a climate situation in the past to be a detailed analogue of the likely climate in the next century with increased greenhouse gas concentrations, it is necessary for the forcing factors (e.g. greenhouse gases, orbital variations) and the boundary conditions (e.g. ice coverage, topography, etc.) to be similar.

Working Group I therefore did not accept that using analogues was an appropriate method for deriving patterns of climate change due solely to greenhouse gas warming.[12] Although this was not a matter for Working Group II to deal with, its final report still included references to scenarios based on palaeo-climatic analogues, but Working Group I's conclusions were the decisive ones.

Budyko also put forward the view that warming might be beneficial to countries at northerly latitudes and that therefore no preventive action was justified. A special meeting with those of the opposite view was arranged in St Petersburg to resolve this issue, but there was still tension at the final meeting of the working group, at which the report was to be accepted. It was, however, obviously important to recognise that in some regards positive consequences of global warming might be expected, particularly at high latitudes, but it was equally important to maintain the principle that the IPCC should not express views on the climate policy to be chosen.

Unease grew also in the course of the work by Working Group II because of the lack of balance in the choice of lead authors of the different chapters in the report. Several scientists from the USSR led the assessment process and scientific as well as practical difficulties arose. This was finally resolved when one of the cochairmen of the working group, M. Tegart, Australia, together with some of his Australian colleagues took on the task of compiling and editing the final report.

6.3.3 Working Group III. Response strategies

The greatest difficulties were, however, encountered in the finalisation of the report of Working Group III.[13] The prime reason was that the IPCC had not emphasised adequately the necessity of distinguishing between factual information provided by scientific research and political judgments in later political negotiations. This became very obvious when dealing with 'Legal and institutional mechanisms'. The task for the IPCC was not to present a proposal for a convention. This section of the report was therefore finally turned into an inventory of issues to be considered in the political negotiations. The summary for policy makers contains a number of questions that were raised without answers being provided. Presumably, this became valuable in the negotiations that began after the acceptance of the IPCC report in the UN General Assembly at the end of the year. It was a necessity to proceed in this way in order to avoid politicising the IPCC work.

The Working Group III report contains a set of four scenarios for future greenhouse gas emissions that show the expected changes from 1985 to the end of the twenty-first century. However, the construction of emissions scenarios for greenhouse gases had not yet attracted much attention from the scientific community. Also, most analyses followed the traditional approach that had been used by the energy industry for decades, i.e. the aim was to provide an analysis of the future supply of fossil fuels that it seemed reasonable to aim for in a continuously expanding market. This essentially meant an extrapolation of prevailing emissions, i.e. an increase of annual emissions from fossil fuel burning by a given percentage per year, in spite of the fact that similar extrapolations had systematically led to overestimates in the past. In fact, the increase had been about 4% per year until the first energy crises in 1973, but had decreased markedly since then. I objected to the approach at one of the last meetings of the subgroup of Working Group III that dealt with this issue, but it was too late in the ongoing race to meet the deadline for the first IPCC report set by the UN General Assembly for any changes to be made.

The so-called reference scenario (A), 2030 High Emissions (Business-as-Usual) implied that the annual emissions of greenhouse gases were projected to increase by about 140% from 5.15 Gt C in 1985 to 12.6 Gt FC in 2030, i.e. approximately 2% per year. (In reality, the annual increase of emissions due to fossil fuel burning during the 20 years 1985–2005 was about 1.8%). Three other scenarios were constructed (B, C and D) assuming different stringent reductions of the high emissions scenario. Deforestation was also assumed to increase, causing an increase of emissions from 1. 7 to 2.6 Gt C per year, without regard for a possible terrestrial sink due to regrowth of secondary forests in deforested areas or enhanced photosynthesis in a carbon-dioxide-rich atmosphere.

Scenario A was used by Working Group I when projecting a global temperature increase of 1.5–3.0 °C by 2030 and 1.5–6.0 °C at the end of the twenty-first century. It is interesting to note that although the projection for 2030 will probably not be realised, the range of the temperature change by 2100 is just about the range of uncertainty of the projected temperature changes as given in TAR, which was completed in 2001. However, TAR also includes the range of uncertainty of the model's sensitivity to a changing radiative forcing. In retrospect I note that the projection of greenhouse gas forcing in FAR was above what was later judged to be a reasonable reference scenario, but not markedly so for the period 1985–2005.

It is to be noted, however that the socio-economic development in the world is largely unpredictable, except with regard to some overarching aspects. For example, the emissions in East European countries and the former USSR decreased by about 30% as a result of the rather chaotic economic and political conditions that prevailed during most of the 1990s. This was not foreseen, even just a year or two before it happened. European emissions have increased less than projected largely because of some concerted efforts, while US emissions have increased about as projected in the late 1980s, and developing countries similarly so.

Thus a rather wide set of scenarios would be required in order to provide a range of alternative future emission paths for the global society, and the associated future changes of climate should not be considered to be predictions, but rather viewed as sensitivity analyses to provide a plausible range of future changes of climate. This distinction between predictions and scenarios is still not adequately appreciated and has sometimes misled the general public, journalists and politicians, which in turn has caused unnecessary controversies. The word 'prediction' should indeed be used with great care. This refers in particular to climate modellers who often view the forcing scenarios as predictions. This distinction was not adequately considered in the report.

As will be seen, however, these IPCC scenarios were later judged to be inadequate, particularly because of the overestimate of the future concentrations of the CFC gases, but also because any projection about the future easily becomes politically contentious.

6.4 The acceptance and approval of the IPCC First Assessment Report

I found it unsatisfactory not to attempt to integrate the findings of the three working groups into a synthesis report and took on the task of writing such a document.[14] I called the IPCC vice-chairman, the rapporteur and the chairmen and vice-chairmen of the three working groups to a meeting a few weeks after the completion of the reports of the three working group in June 1990 to discuss a draft that I had prepared. It was intended to be the prime document for consideration by the IPCC at its fourth plenary session in Sundsvall, Sweden, in late August. The plenary sessions of the three working groups had approved the summaries for policy makers in their respective sessions. They had therefore merely to be accepted in the IPCC plenary session. This should have left time for a more penetrating discussion of a synthesis document. Major modifications were, however, proposed in the course of the plenary session and the negotiations came close to a breakdown. In retrospect it is easy to understand how this difficulty arose when such a key document was not circulated in advance with a request for written comments for a modified draft that would be finalised at the plenary session.

This fourth IPCC session became the meeting place for, on the one hand, scientists that had put forward a factual evaluation as supplied by the IPCC working groups and, on the other, government representatives with the responsibility of ascertaining if the report was structured in a way that would serve the domestic political process in their respective countries. It became particularly difficult to deal with the sections on response strategies. Shorter versions of the three summaries for policy makers from the working groups were finally harmonised and brought together into one document. It was not a synthesis in the true sense of the word, but it was approved after a long final session that ran well into the morning of the following day.

A report from the working group on the participation of developing countries also became part of the final assessment report. It emphasised the difficulties experienced by developing countries in taking part in and contributing to the work of the IPCC. Their human and financial resources were limited and they often lacked the kind of institutions that were required to analyse the climate change issue in detail. Clearly, seen from their perspective more attention needed to be given to this in the future. It became difficult, however, to include what was proposed this late in the process. This triggered some of the most heated discussions of the synthesis report. Brazil and Mexico were in the forefront of trying to safeguard the interests of developing countries as well as their own. The final report was changed to point out that

. . . as the greenhouse gas emissions in developing countries are increasing with their population and economic growth, rapid transfer, on a preferential basis to developing countries, of technologies which help to monitor, limit or adapt to climate change, without hindering their economic development, is an urgent requirement . . . Recognising the poverty that prevails among the populations of developing countries, it is natural that they give priority to achieving economic growth. Narrowing the gap between the industrialised and developing world would provide a basis for a full partnership of all nations in the world and would assist developing countries in dealing with the climate change issue.

These sentences are of course politically inspired, but they are basically factual They were, of course, expressions of the frustration that delegates from developing countries had often felt in carrying out the assessment. They would have to be considered carefully when planning the next assessment. The fact that an agreement was still reached signalled that there was a common spirit amongst the participants and a recognition that the climate change issue had to be taken seriously, but the problem of how to conduct the forthcoming sessions of the IPCC remained. This recognition of developing countries' concerns was, however, an achievement that would become very important in the coming years and the scientific conclusions were not compromised because of these, in a sense, political interventions. On the other hand, the much more difficult problem of financially and technically supporting developing countries in their efforts to eradicate poverty and raise their living standards as required in order for them to address the climate change issue is, of course, a political problem far outside the IPCC's responsibility.

It is finally interesting to note that this synthesis report and its conclusions were seldom explicitly referred to in later years. Money for its publication was not in the IPCC budget and the full report did not appear in print until June 1992, and was then paid for by a dozen countries as a special contribution to the IPCC Trust Fund. The working group reports, on the other hand, and particularly the one by Working Group I, were seen as important sources for scientific and technical information about the climate change issue. Their publication was quickly taken care of by the UK signing an agreement with a commercial publisher (Cambridge University Press, Working Group I Report), and through special arrangements by the governments of Australia (Working Group II Report) and the USA (Working Group III Report), i.e. by the countries that had carried the prime technical and financial responsibility for their completion.

6.5 Scientific input in the negotiations about a framework convention

The IPCC FAR was submitted to the UN General Assembly in October 1990, but before it was dealt with at the assembly it was also considered by the Second Climate Conference that was arranged in Geneva at the invitation of the Secretary General of WMO. Another appeal for action and also a call for intensified research was agreed. I was not involved much with this conference and actually found it somewhat superfluous, because not much could be added to the IPCC report and decisions would anyhow have to be taken by the UN General Assembly. It was, however, presumably of some importance as a preparation of the later formal handling of the issue in New York.[15]

The report was discussed at the UN in December and it was decided that an Intergovernmental Negotiating Committee (INC) be formed. The INC was given the task of preparing for the submission of a proposal for a framework convention on climate change (FCCC) to the UN Conference on the Environment and Development (UNCED), scheduled for Rio de Janeiro in June 1992.[16] This meant that only 18 months were allowed for the completion of this by no means simple task.

The questions then also arose: What will the future role of the IPCC be? Would there be a place for independent scientific assessments in the future or would these be the responsibility of a convention? It was interesting to note, however, that while the INC was charged with preparing and negotiating a climate convention, the question of how to continue the analysis of the progress in the field of science for the time being was left for the IPCC. At that moment it was essential that the IPCC was an intergovernmental organisation. I therefore simply took it for granted that its work would continue after an agreement on the creation of a climate convention had been reached, and so it finally did.

The INC met near Washington DC, USA, in February 1991. The earlier invitation by President Bush to host the meeting had been accepted. Jean Ripert of France was elected chairman and as usual in the UN much time at this first meeting was spent on agreeing a structure for conducting the work and on elections of chairmen and other officers of the committees that were created. More than 100 people from the UN in New York attended the transformation of the climate change issue into a central UN effort. I really wondered what they were doing, since their participation in the work so far had been minimal.

Seen from an IPCC perspective, however, the crucial issue of how the IPCC could best contribute to the work of the INC remained. It was of course significant that Jean Ripert, chairman of the INC, also had been a French delegate to the IPCC. He knew well its work so far. As chairman of the IPCC I had become well acquainted with him.

Others of those that had previously been delegates to the IPCC now represented their countries during these political negotiations. Further, it was recognised early that the work of the INC should be based on the IPCC assessments and it was decided to issue a standing invitation to the IPCC chairman to represent the IPCC at INC sessions. In fact, I attended all but one of the eleven meetings that the INC held before the FCCC came into force in 1994. This, I felt, was most important in order to keep the INC delegates well informed of the progress that had been made in the work towards new scientific assessments, and equally to keep the IPCC work apart from the political negotiations as well as possible.

The IPCC held its fifth meeting in March 1991 in Geneva.[17] On the basis of the experience gained during the first 2½ years of its existence the session agreed formally a set of rules for the conduct of its work: *Principles Governing IPCC Work*.[18] It is to be noted that these rules still provided great freedom for the Bureau and the chairman of the IPCC in conducting the IPCC work. In particular, the authority to invite scientists to carry out the assessments was given to the IPCC Bureau, although governments were asked to nominate candidates and were informed about the choices made. This was a very important decision that aimed at preserving the scientific integrity of the assessment process. It was also made clear in these rules that different views that might arise in the course of the assessment should be explained and recorded, i.e. unanimity was not required. Elisabeth Dowdeswell of Canada played an important role in codifying the procedures as they had evolved since the creation of the IPCC, with due regard for the WMO rules that she knew well from having been a member of the WMO executive council for a good number of years. It is quite remarkable that the rules of procedure were simply a dozen short paragraphs. This undoubtedly implied that the IPCC had gained the confidence of the countries that had played a leading role in the IPCC so far. This was going to be most valuable in the future.

There were also constructive discussions about a work programme for 1991 and beyond. Based on a judgement of the issues that might be of prime importance for the INC in its task of formulating a text for a climate convention, I proposed to the members of the IPCC Bureau that the following topics be pursued during the next year:[19]

1. methodologies to be used for building a data bank for national net greenhouse emissions;
2. projections of the regional distribution of climate change and associated impact studies, including model validation studies;
3. issues related to energy and industry;
4. forestry-related issues;
5. selected studies of the economic implications of climate change;
6. vulnerability to sea level rise;
7. emissions scenarios.

I had also indicated to the INC at its first meeting in February 1991 that these issues probably would be given priority by the IPCC in its pursuit of a special assessment as a contribution to the Rio conference. It should be noted that the assessment of the detailed progress in the field of climate research was not emphasised, but implicitly assumed to be dealt with, but progress of this kind was not of direct political interest and thus was not the focus in the contacts with the INC. It should be recalled, however, that the procedures introduced by the IPCC in order to secure scientific acceptance of its work also meant that the opportunities for producing supplementary reports were limited. Nevertheless the IPCC considered it desirable to prepare one focussing on

1. greenhouse gases, their sources and sinks and a revision of the projections of future greenhouse gas concentrations in the atmosphere;
2. climate modelling, prediction and model validation; and
3. observed climate variability and change.

The goal was to publish this supplementary report to the Working Group I Report on the Scientific Assessment 1990 in time for the UNCED meeting in Rio in June 1992.[20] This was indeed a bold attempt. There were merely 15 months before UNCED opened in Rio. Analyses of other issues were also to begin, but were rather be viewed as preparations for a more complete Second Assessment Report (SAR), which at that time was scheduled for completion in 1994 or 1995. A supplementary report on the 1990 impacts assessment also appeared in 1992.[21]

The IPCC further requested the Organisation for Economic Cooperation and Development (OECD), in cooperation with the International Energy Agency, IEA, to assist in the improvement of methodologies for assessing greenhouse gas emissions. This was to be carried out under the guidance of a steering group set up by the IPCC Working Group I, and pursued in accordance with the work plans adopted at the meeting. This led to a close cooperation between the IPCC and the OECD on these matters until the Climate Convention came into force and joint responsibility for this important activity was accepted by the two organisations. This was a major undertaking. It required substantial financial resources beyond what was available through the IPCC Trust Fund and would not have been possible to complete without the substantial contributions made available through the OECD and later also the INC.

It was also decided to intensify the efforts to deal with issues related to energy and industry in a long-term perspective. Presumably because of the recognition of its work by the UN, a large number of organisations approached the IPCC offering their services to assist, and presumably in this way to become players, in the further pursuit of this very important issue.

The IPCC gave priority to the initiation of country studies. The discussions showed quickly that countries were anxious to conduct such studies to become familiar with the climate change issue. The Global Environmental Facility (GEF) at the World Bank and the USA were willing to provide substantial financial support for such work.

The discussion of the possible economic implications of climate change was not completed at the session. In view of its obvious importance and because of comments made during the discussions, I took the initiative of setting aside some time at the next IPCC session for discussions of this matter, and invited a few key economists to address the IPCC.[22] Although it was not possible to present this issue at the ongoing negotiations about a Climate Convention, steps had thus already been taken in 1991 to familiarise IPCC delegates with this issue, particularly those from developing countries. The four people that were invited were to become much engaged in the climate change issue during coming years.

The UN decision in 1990 was also noted by industry, above all by corporations responsible for supply of energy. I noticed this changing attitude because of receiving invitations to present the IPCC findings. In July 1991 the International Gas Union held one of its regular meetings in Berlin and the Thirteenth World Petroleum Congress met in Buenos Aires for its triennial meeting in October that same year. I addressed both conferences.

The theme for the International Gas Union Congress was 'Natural Gas, the Clean Energy'. The emissions of carbon dioxide as a by-product when burning gas were largely ignored. The comparatively modest emissions of carbon dioxide from producing a given amount of energy when using natural gas as a fuel rather than oil or coal could of course have been emphasised, but it was obviously at this stage not advantageous to associate the use of natural gas with any kind of air pollution. My presentation was attended by only a few of the participants in the congress and I doubt that my presentation had any impact at all.[23]

I met a similar attitude at the World Petroleum Congress in Buenos Aires in October that same year.[24] On this occasion Dr W. Nierenberg, former Director of Scripps Oceanographic Institute, La Jolla, CA, USA had been invited as a critic of the IPCC conclusions in its FAR. He also represented the Marshal Institute, a newly formed lobby organisation in the USA. Nierenberg's introductory statement was an acceptance of some principal points that had been presented by the IPCC, i.e. that '. . . a doubling of the atmospheric carbon dioxide concentration in fifty to one hundred years might be a grave matter . . .' He referred, however, to the chaotic climate system and the limitations of its predictability and did not expect an increase of the global mean temperature by more than 0.5 °C, partly because the additional absorption of thermal radiation decreased as the carbon dioxide concentration increased.[25] This was in itself a correct notion that had been considered in the IPCC analyses. He systematically underestimated the observed change so far and concluded his analysis of likely future changes by a very simple 'back of the envelope' computation that was simply wrong.

Towards the end of our encounter Nierenberg asked rather abruptly for our discussions to be concluded. I do not know if this encounter should be given any significance, but it was obvious that few, if any, initiatives would be taken during coming years by those being engaged in the supply of fossil fuels as a primary energy source. The participants of the Congress took little notice of this debate and the Swedish National Committee of the organisation was somewhat irritated at having to pay for my attendance in Buenos Aires. There were also some other organised objections to the IPCC assessment. Thus Fred Singer at the University of Virginia published a book with contributions from a of number of people sceptical of the IPCC's efforts (Singer, 1992). His past professional record undoubtedly gave him a platform to reach out from. There were, however, very critical views amongst the key experts about his way of dealing with the issue. He was no longer an active researcher who contributed to the scientific debate in the professional literature, but appeared frequently in the daily press. A few of his remarks about statements in the policymakers' summary of the IPCC FAR which were published in the book, might illustrate his approach.

He quoted from the assessment: '. . . It is not possible to attribute all, or even a large part, of the observed global-mean warming to the enhanced greenhouse effect, . . .' and claimed that this conclusion was not reflected in the policymakers' summary. This is simply wrong. In fact the IPCC says: '. . . The size of this warming is broadly consistent with predictions of climate models, but it is also of the same magnitude as natural variability. Thus the observed increase could be largely due to this natural variability and other human factors . . .'

It is incorrect when Singer said '. . . the [Policymakers] Summary is essentially a document of the governments, not the scientists . . .' The fact is that it was prepared by scientists and approved in the presence of the lead authors of the main report and government representatives, many of whom were scientists. It was actually not possible to include statements that on scientific grounds were not approved by the lead authors.

Singer considered it to be misleading to say that '. . . we are certain that . . . there is a natural greenhouse effect . . .' because it gave the impression that this was a newly discovered finding, which it was not. The IPCC, on the other hand, meant that it is important to point out that human activities are modifying a natural process that is of major importance for the existence of life on earth, which of course is a fundamental fact.

These intentional misinterpretations of the IPCC report were annoying, and almost all scientists that I have met considered Singer's activities during the early 1990s, and actually ever since, to be a systematic attempt to discredit the IPCC's efforts by making incorrect or misleading statements both verbally and in the popular press. On the other hand, some of the articles that were included in the book raised legitimate questions and accordingly were considered in later IPCC assessments.

The final meetings of the IPCC working groups II and III before the UNCED in Rio were held in January–February 1992, while Working Group I met in

Guangzhou in China. The prime issue then was the acceptance of the different chapters of 'The 1992 IPCC Supplement: Scientific Assessment' and the formal approval of its summary for policy makers (IPCC, 1992b). The discussions about this report were intense and disagreements were not easy to resolve. A crucial chapter was the one presenting a set of six new scenarios for greenhouse gas emissions during the twenty-first century and especially the expected increases of atmospheric carbon dioxide and other greenhouse gas concentrations. The discussion had thus already returned to the issue of scenarios for the future, socio-economic issues that govern the industrial development, and the way an increasing future need for adequate energy supply would be met. The key words are of course *adequate energy supply*. What does this really mean? It is not easily resolved.

Time had been short and work began late. The specifications of the assumptions made in order to project what might happen more than 100 years into the future were simple but, many thought they were not acceptable. They implied a wide range of possibilities, but the basic documentation for the analysis was not available at the meeting. For example, it was difficult for many to accept that the world population in 2100 might be anything between 6.4 and 17.6 billion people and that the world gross domestic product might grow 20-fold, which still was substantially less than the value projected by the World Bank. Another major uncertainty was the magnitude of the global fossil fuels reserves and resources which were kept secret by many oil and gas companies. All these became major obstacles in gaining acceptance of the chapter and understandably so. Non-governmental representatives, in this case particularly those representing the energy industry, objected to it very strongly.[26]

The report had, however, been subject to an extensive review, and amended accordingly. The reasons for the objections were to a considerable extent political and came from special interest groups, but there was also genuine uncertainty about likely future emissions. The incident showed the difficulties encountered when the underlying data and analyses were not available well in advance. It also became clear that the polarisation of opinions about the seriousness of the climate change issue was increasing. The report was finally accepted with some modifications and full explanations of the assumptions having been made, the scenarios IS1992 a, b, 'business-as-usual'; IS1992 c, d, very low, and low emissions; and IS1992 e, f, very high, and high emissions, were accepted. There were simply no better ones available at the time. Actually the scenarios spanned a wide range and could well be used for analysis of the sensitivity of the climate system to alternative emissions scenarios, but should of course not be seen as predictions for the future. They would in that sense be useful for quite some time to come. The issue was, however, soon to be raised again by representatives of the energy industry.

The objections from some stakeholders were summarised in a letter from the Climate Council, the Edison Electric Institute and the National Coal Association to the chairman of Working Group I and myself just as the supplementary report appeared a few weeks before the UNCED opened in Rio. The subject matter itself was not touched upon, but the four objections listed all concerned procedural issues with a request for these to be brought up specifically at the next session of the IPCC. The difference between the scientific, factual approach, on one hand, and this thinking by lawyers and industrial representatives defending special interests, on the other, was really no surprise.

Since the other working groups were not trying to present additional information to the last session of the INC before Rio, their meetings were not controversial. It is of interest, however, to quote a few sentences from the preface of the report on impacts:

... if carbon dioxide in the atmosphere doubled, there would not be a global catastrophe due to climate change. However, there would be severe impacts in those regions of the world least able to adapt and substantial response measures would need to be taken.

The key findings in the new IPCC assessment were presented to the INC early in February 1992, but it was clear that by then the negotiations had become purely political and there was no time to present new scientific information. On the other hand, the fact that scientific arguments were not the focus of attention could also be seen as progress in that the IPCC FAR had largely been accepted and was thus serving its purpose. The IPCC presentations to the INC had been important in order to maintain the flow of scientific information and to keep the picture of a possibly changing climate in focus.

It was, however, not possible to resolve the political controversies at this INC session. The USA kept strictly to the simple view of *no targets and no timetables*. The convention could indeed only be a framework convention. In the final negotiations, the approach of the UNCED in Rio put pressure on countries. The final text did contain some more specific goals that would be of considerable importance during the following years, but few articles went beyond just providing a framework for continued negotiations. The INC chairman called for an additional meeting in April and some further compromises paved the way for an agreement.

I will limit myself to quoting and commenting on some parts of the Convention that were of particular interest for the further work by the IPCC:[27]

Article 2: Objectives
The ultimate objective of this Convention and any related legal instruments that the Conference of the Parties may adopt is to achieve, in accordance with the relevant provisions of the Convention, stabilisation of greenhouse gas concentrations in the atmosphere at a level that would prevent dangerous anthropogenic interference with

the climate system. Such a level should be achieved within a time frame sufficient to allow ecosystems to adapt naturally to climate change, to ensure that food production is not threatened and to enable economic development to proceed in a sustainable manner . . .

The interpretation of the expression: *dangerous anthropogenic interference* is of course a political issue and the precise role of the scientific community in this context is therefore unclear, as will also be seen in the course of the later work by the IPCC.

Article 4: Commitments
1. All Parties, taking into account their common but differentiated responsibilities and their specific national and regional development priorities, objectives and circumstances, shall: . . .

A number of general commitments are then enumerated about reporting, programmes for mitigation, management, cooperation, etc, but they are not specific and therefore not binding with regard to measures aimed at slowing down an expected climate change. The article then continues

1. The developed country Parties and other Parties included in Annex 1 commit themselves specifically as provided in the following:[28]
    a. Each of these Parties shall adopt policies and take corresponding measures on the mitigation of climate change, by limiting anthropogenic emissions of greenhouse gases and protecting and enhancing its greenhouse gas sinks and reservoirs. These policies and measures will demonstrate that developed countries are taking the lead in modifying longer-term trends in anthropogenic emissions consistent with the objective of the Convention, recognising that the return by the end of the present decade to earlier levels of anthropogenic emissions of carbon dioxide and other greenhouse gases not controlled by the Montreal Protocol would contribute to such modification, and taking into account the differences in these Parties' starting points and approaches, economic structures and resource bases, the need to maintain strong and sustainable economic growth, available technologies and other individual circumstances, as well as the need for equitable and appropriate contributions by each of these Parties to the global effort regarding that objective.

In the following subparagraph it is specified that the aim of the parties should be to return

. . . individually or jointly to their 1990 levels [of] these anthropogenic emissions of carbon dioxide and other greenhouse gases not controlled by the Montreal Protocol.

It was very obvious in the course of the negotiations how countries attempted to safeguard themselves in order not to be victims of more far-reaching commitments, which of course is a normal feature in the negotiations of international agreements. It was clear, however, that the overarching, long-term implications of a global climate change were not yet well appreciated.

The agreed text of the Convention was submitted to the Rio conference a month later and was signed by 156 countries. Ratification by 50 countries was required before the Convention would come into force, a rather modest requirement.

The achievement was celebrated at a special ceremony. I found it rather strange that one basic reason for the success was the thorough scientific analysis of the issue by the IPCC, but this seemed to have been forgotten. Not until the celebrations were in full swing did the INC Chairman, Jean Ripert, informally call me to the podium to join the others who were congratulating each other. The WMO and UNEP were apparently not eager to recognise formally the importance of the efforts by the IPCC and thereby the scientific community. I interpreted this incident as an indication that the IPCC, also an intergovernmental organisation and thus rather independent, was viewed as a threat to the more traditional procedures within the two parent organisations of the IPCC, i.e. the UNEP and the WMO.

6.6 What has experience so far to say about the role of science?

First of all, it does not seem likely that a Climate Convention would have been agreed at Rio if a well-organised and scientifically credible assessment had not been available in 1990. This assessment in turn was possible only because assessments initiated by the US NAS and the international scientific community had begun a decade earlier. The emergence of the climate change issue was primarily scientifically driven.

Our knowledge about possible future changes of climate was in many regards still limited and uncertain when the IPCC First Assessment was presented to the UN General Assembly in 1990. This was clearly pointed out by the IPCC. A less carefully prepared assessment could have been seriously questioned, but analyses by well-recognised scientists gave credibility to the reports particularly that of Working Group I on the scientific basis, and these could therefore serve as the basis for negotiations. Also, country delegates to the INC could ask scientists at home about their views and receive an independent scientific evaluation, since hundreds of scientists from all over the world had taken part in the IPCC work and had in a way implicitly agreed with the conclusions.

The direct contacts between negotiators and representatives of the IPCC also meant that a mutual respect had gradually been established between these groups. It was obviously important not to hide uncertainty, and also to stress the implications of the inertia of the climate system. The likelihood that climate change might be with us for many decades, perhaps a century or even more, once it began, seemed to have been a forceful argument, and rightly so.

'Global warming' quickly engaged environmental groups. Greenpeace, in particular, was active early and published a detailed (paper-back) report in 1990, having recruited many well-known scientists as contributors (see Leggett (1990)). Greenpeace's arguments in the public debate, on the other hand, were not always scientifically well founded. I told the leader of Greenpeace my view about this mismatch between their scientific analyses and action strategy, but with no satisfactory response. Greenpeace had a significant influence on delegates, not least those from developing countries. In fact, the pronouncements by non-governmental environmental organisations were quoted more often in the plenary sessions than the IPCC reports. Their interventions in the political debate were, however, largely based on work by the IPCC. National representatives recognised that they needed an authoritative basis for their activities.

It was of course legitimate at the time to ask the question: how much more definite would the scientific findings have to be before calls for action were justified? It was obvious that no coordinated action would be agreed before the Convention had been ratified and a first session of its parties had taken place. Not only did the climate system have considerable inertia, but so also did the global socio-economic system of the countries of the world and their representatives, i.e. the politicians. I am sure, however, that few imagined that ten years later a protocol specifying more precise measures with binding commitments would still not have come into force. The Kyoto Protocol agreed in 1997 came into force and became part of the Convention in 2005, i.e. eight years after its acceptance in Kyoto in 1997. There have later been claims that the scientific community primarily acted in self-interest, i.e. the greater the scare about a human-induced climate change the greater the resources that would become available for scientific research (see Boehmer-Christiansen (1994a,b)). The analysis above should have made it clear that an honest engagement by the scientific community was a prerequisite for a successful completion of the FAR. A more thorough analysis of this particular issue is, however, given in following chapters.

7

Serving the Intergovernmental Negotiating Committee

The IPCC is reorganised and begins another scientific assessment; an optimistic attitude still prevails, although sceptics increase their objections.

7.1 Changes in the IPCC structure and new members of the Bureau

The IPCC had implicitly been given a clear task by the large number of countries that signed the FCCC in Rio: i.e. to continue the assessments and serve the INC in its work towards a first meeting with the parties of the Convention that might take place within the next few years. Signatures from only 50 countries were needed for the convention to come into force. Our understanding of the environmental aspects of the issue now needed to be broadened and a more penetrating assessment was needed of the impacts of a climate change and its associated costs, together with socio-economic studies of the implications of mitigation and adaptation. But whenever value judgements were an essential part of such issues, care had to be exercised in order to avoid criticism of the IPCC efforts because of possible implicit political assumptions in the course of the analyses.

As was pointed out in Section 6.5, it was for the IPCC to judge how the expression *dangerous interference with the climate system* should be interpreted, but it was of course still important that relevant information was provided to serve as a basis for political analyses and negotiations. This was more important now, when industries were gradually becoming aware of the possible threats to their activities, which was how many of their representatives viewed the emerging situation. A rather hostile attitude towards the IPCC was developing because of the claims that changes of climate in the long term might imply a need for substantial changes of the future world energy supply systems.

In the course of pursuing preparatory work for UNCED in Rio it had become clear that some major changes in the structure of the IPCC were needed, and also in the way its major tasks were undertaken. This became a more important issue during the eighth IPCC plenary session which was held in Harare, Zimbabwe (see IPCC (1992d)). The IPCC had by then been firmly established in the UN system and had gained a crucial position with regard to the politics of the climate change issue. A climate convention might be a reality within a few years. How could the IPCC best contribute in this context and what would its future role be? It was by no means self-evident how to settle the complex issue of the interaction between science and politics. Actually a close interaction was very worrisome to many scientists, although the majority were anxious to contribute in a constructive manner, as long as the integrity of the scientific process was safeguarded.

The report by *Working Group I* on the science of climate change had largely been well received by both the scientific community and the media and politicians. Work towards a second assessment could therefore proceed in a similar manner, but a few additional tasks seemed essential:

- Methodologies for the preparation of national inventories of anthropogenic emissions by sources of and removal by sinks of greenhouse gases had to be further developed in a cooperative effort between IPCC member countries, the OECD and other organisations as appropriate. Guidelines for intercomparison of different methodologies should be agreed, taking into account the requirements of the Convention.
- On this basis and by using other means, the best possible assessment of past and present national and global net emissions of greenhouse gases, direct as well as indirect, should be made.

It should be emphasised that the early initiatives taken by the IPCC to prepare for a systematic gathering of information regarding sources and sinks of greenhouse gases and assessments of stocks of greenhouse gases in the atmosphere were essential. This was largely a technical matter and meant the establishment of links with national environmental agencies. The IPCC aim was to prepare global inventories as soon as possible to assist future assessments and evaluations. The best possible basis for the estimates of likely future changes of greenhouse gas concentrations in the atmosphere would thus gradually become available.

It was also clear that the cooperation between Working Groups II and III had to be extended. It was agreed that a new *Working Group II* should assess available scientific, technical, environmental, social and economic information regarding impacts of climate change as well as response options to mitigate and adapt to climate change. This task was obviously huge and the work was therefore further subdivided and given to four subgroups, A, B, C and D.

> *Subgroup A* dealt with issues of energy, industry, transportation, air quality and health, waste management and disposal.

Subgroup B dealt with issues concerning small islands and coastal zones, oceans and marine ecosystems, tropical cyclones, storm surges and sea level change.

Subgroup C dealt with issues concerning unmanaged resources and terrestrial ecosystems, mountain regions, the cryosphere (i.e. ice and snow), hydrology and terrestrial impacts of climate change, for example due to floods.

Subgroup D dealt with issues concerning desertification, droughts, agriculture, forests, land use (including human settlements), health management and water resources.

Further, it was emphasised that the study of impacts and policy options would best be done in the broader perspective of sustainable development. The time was, however, not yet quite ripe for globally integrated assessments.

It was also decided to create a new *Working Group III* to deal with crosscutting issues related to climate change. I had taken the initiative of devoting the good part of a day at the sixth IPCC plenary session (October 1991) to presentations of socio-economic issues that might arise in combating human-induced climate change.[1] It was obvious that there was rapidly increasing interest in the field of economics. The development of macroeconomic models for more internally consistent scenarios of future emissions of greenhouse gases under various assumptions about key factors might become increasingly important in forthcoming assessments. The implications of the six scenarios that were presented in the supplementary report on climate change (IPCC, 1992b), needed to be studied further and compared with other efforts of a similar kind. The costs of alternative mitigation efforts had also become an important political topic. Developing countries might also raise issues about the lack of equity in the world in future negotiations. In my opening statement to the eighth session of the IPCC (see IPCC (1992d), Appendix D), I spent considerable time outlining key objectives for such work by the IPCC, building on the seminar that had been held in 1991. The precise formulation of the task led, however, to considerable controversies. It became very obvious how closely politics and scientific research in the field of economics were tied together. Two topics were finally agreed and the working group was given the opportunity of adding further topics to its agenda in the course of its work, after approval by the IPCC. Difficulties would, however, emerge later. The prime tasks at the time were:

- Assessment of the socio-economic consequences of adaptation to and mitigation of climate change in the short- and long-term perspective at regional and global levels.
- Development of a range of internally consistent scenarios for future emissions based on reasonable economic, demographic and technological projections, taking into account gaps and uncertainties in available knowledge.

It was once more specifically emphasised that the working groups should include a carefully conducted peer-review process in the preparations of their reports. This was to be done in two steps, first by asking the opinions of individual scientists in the field, and second by requesting government institutions and agencies to nominate reviewers, who would bring the views of government experts into the review process. This may seem like letting political considerations influence the reports that had been written by teams of scientists and other experts. It should, however, rather be viewed as an attempt to broaden further the interplay between scientists and governments. It was prescribed that modifications of the drafts of the reports should be done with reference to scientific papers and/or reports that had appeared in peer-reviewed scientific journals and other well-established publications.

The procedures for writing and approving the IPCC report had to be applied precisely and stringently. The detailed bulk reports and the technical summaries were not to be formally approved by the plenary sessions of the working groups but merely accepted as fair summaries of present knowledge and they were to be published under the names of their lead authors. On the other hand, the working groups were to formally approve the summaries for policy makers at plenary sessions. Proposed modifications of the summaries drafted by the lead authors and the working group bureaux had to be based on the underlying bulk reports and accepted by the lead authors as such. Additional references could be brought into the process in support of proposed modifications, if so requested by a reviewer. In fact, the inclusion of technical summaries as part of the bulk reports was decided upon because of the difficulties that had been encountered when seeking approval of the summaries for policy makers in the working groups, as these summaries had become rather extensive. They might then be more concise, since details could be included in these technical summaries. This somewhat cumbersome procedure later turned out to be very important although sometimes still controversial.

The principles for the election of members of the IPCC and the working group bureaux were also somewhat modified. I had myself already been reelected as the IPCC chairman for four years at the sixth session of the IPCC (in October 1991), but all other members had to be appointed anew with the condition imposed that no country could have more than one member on the Bureau. It also became clear that a better balance between delegates from developed and developing countries had to be achieved. I proposed, and it was accepted, that two cochairmen, one from a developed and one from a developing country, would lead each working group, and rather than having one IPCC vice-chairman and a rapporteur I proposed that there should be two vice-chairmen. The principle had been established earlier that an elected vice-chairman would not automatically replace the chairman, if the chairman left his post before his term expired. Instead, a new election had to be called.

Cochairmen and vice-chairmen were elected for each of the three working groups and all members of the working group bureaux became members of the IPCC Bureau. Finally, one delegate from each of the six WMO regions was chosen in order to incorporate a feature of the WMO structure. The IPCC Bureau thus consisted of the IPCC chairman and representatives from altogether 27 countries. A developed country that aspired to the cochairmanship of a working group had to supply the means of maintaining a technical support unit to serve the working group. The election of the new members of the Bureau was not simple because of the many wishes and aspirations that were expressed by country representatives, but it was finally achieved by consensus.[2] No nomination committee had been appointed and I felt strongly the support of delegations in the course of this process.

Dr Al-Gain from Saudi Arabia continued as vice-chairman of the IPCC and Yuri Izrael of the Russian Federation became the other vice-chairman. This move of Dr Izrael from the cochairmanship of Working Group II was a reflection of the difficulties encountered in the course of the work for the FAR. It became feasible because of the collapse of the USSR and its replacement by the more loosely structured Russian Federation. There were no objections to these changes from the Russian Federation.

The UK continued its key role of chairing *Working Group I* with Sir John Houghton as cochairman and with Gylvan Meira Filho from Brazil as the other cochairman.

The developed country chairmanship of *Working Group II* was turned over to the USA. Robert Reinstein of the USA had replaced Dr F. Bernthal in 1990 when the USA had held the chairmanship of Working Group III. Now he took charge of Working Group II together with a cochairman from Zimbabwe, Marufu Zinyowera. It is interesting to note that Reinstein also served as the principal US delegate to the INC. The US State Department tried to tie the IPCC and the Climate Convention more closely together. Dr Robert Watson replaced Reinstein as cochairman of Working Group II in 1993 when Bill Clinton became the new US President and Clinton also appointed Watson as assistant science advisor in the White House. Watson had also led the UNEP assessment of the ozone issue, which meant that a very valuable interconnection was established.

Soon after having been elected as cochairman of *Working Group III* Ms E. Dowdeswell from Canada was appointed as executive director of UNEP and was replaced by Dr J. Bruce, Canada, a former director of the Canadian Weather Service. Dr Hoesung Lee, who had a prominent position in the Republic of Korea dealing with energy issues, filled the other post of cochairman. Their expertise in economics was, however, limited.

Continued support from UNEP was now to be expected under Ms Dowdeswell's leadership. On the other hand, it was obvious that the fact that the IPCC was an intergovernmental organisation worried the secretary general of WMO, Patrick Obasi, whose responsibilities at the UN covered meteorology, climatology and hydrology. I had noted this before, but it did not influence the reorganisation of the IPCC.

I had overall responsibility for the work towards the Second IPCC Assessment and primarily based my work on discussions with the six cochairmen of the working groups. I was satisfied that the three from developed countries were very knowledgeable in the environmental sciences and well established in professional circles, while those from developing countries brought other but more limited expertise, and their participation was essential as an educational process. In general, the mixture of representatives from developed and developing countries served the IPCC very well during the next five years.

The three working groups were supported by their respective technical support units. The three men in charge of these were Dr Bruce Callender, UK (Working Group I), Dr Richard Moss, USA (Working Group II) and Dr Eric Haitis, Canada (Working Group III), and they were invaluable in organising the practical details of the assessment work.

There were, however, some complaints about the preponderance of Anglo-Saxons in the leading positions of the IPCC. Nevertheless, I felt that this would not be a disadvantage during the coming years. The fact that English was the working language and that simultaneous translation was provided only at the plenary sessions of the IPCC and the working groups was an obvious disadvantage for many of the delegates. The work of establishing a knowledge base from the available literature was done by small groups of scientists who served as lead authors and was carried out almost exclusively in English. It was essential that scientific papers and reports in other languages were not unintentionally left out. The review process served as a safety valve in this context, although not always adequately.

It is obvious that not all other members of the Bureau were nominated solely as experts, although most of them had considerable experience of scientific and technical work in relevant fields. Some were presumably given political instructions by their respective governments, which to some degree might also have been true for country delegates to the IPCC in general. This was unavoidable in an intergovernmental body and was not necessarily a disadvantage if a stringent distinction was kept between scientific and political arguments in course of the assessments. I was well aware of this, and had to watch it carefully and conduct the assessment process accordingly. The teams of scientists carrying out the assessments (the lead authors) had as far as possible to be chosen on the basis of their scientific and technical competence. The working group bureaux carried the responsibility for this task, which was accomplished in consultation with the members of the core group. Countries nominated lead authors, but the IPCC was free to choose amongst those that were nominated.

So far the IPCC work had been governed by the WMO rules of procedure. It became clear, however, that there might be a need to expand them in order to safeguard the assessment process from political interventions. I was approached by Don Pearlman,[3] at the Working Group I plenary session in Guangzhou to discuss this matter. Lobby organisations were being formed, still primarily in the USA, with financial support from some industrial groups and with the aim of systematically countering the assessments by the IPCC. Don Pearlman had served as an expert in the US Department of Energy during the presidencies of Ronald Reagan and George Bush and was now employed by the coal industry. The Global Climate Coalition, led by John Shlaes, was another similar industrial group. The strategy pursued was primarily to minimise the significance of the possible impacts of climate change and to address procedural and legal issues. To focus on a revision of the IPCC rules of procedure was then an obvious first step and, in fact, Pearlman did have some influence on the modifications of the IPCC rules of procedure for a few years. Controversies between stakeholders, particularly in the energy industry, and the IPCC were still in their early phase in late 1992, but would increase and indeed become troublesome during the coming years.

In retrospect I judge that the IPCC and its Bureau were able to function well. On the whole, decisions were not unduly influenced by politics, but were rather almost entirely scientific and technical. The fact that the IPCC was an intergovernmental body had, on the other hand, the advantage that government delegates and scientists in a sense became partners in the assessment process. Others did not think that this was appropriate as will be seen in the later account of the preparations for the SAR. It is, however, my firm conviction that the scientific integrity of the IPCC was not jeopardised, but constant attention to this matter was of course essential.[4]

7.2 Cooperation with the Intergovernmental Negotiating Committee

The ninth IPCC session was held in Geneva in June 1993 (see IPCC (1993)). Work plans were agreed for the special report on radiative forcing of climate change, and an evaluation of the IPCC IS92 Emission Scenarios to be completed in 1994, and for the IPCC SAR then scheduled for the latter part of 1995. Cooperation with the INC was of course very important and accordingly its chairman, Roul Estrada-Ouyela from Argentina, was invited to the IPCC meeting. He gave a detailed analysis of the INC needs and a valuable cooperative spirit emerged from the discussions. The IPCC had to spell out clearly its timetable for the assessments to be delivered because of the rather lengthy process that was required in order to arrive at reports that were accepted by the scientific community. This meant a balance between a quick response to serve the INC and care in the assessment work.

It was already becoming clear at that time that the Climate Convention might come into force earlier than expected. Altogether 166 countries had by then signed it and 29 ratifications had been recorded at the time of the meeting. Fifty ratifications were required for the Convention to come into force and this was achieved in March 1994. The first conference of the parties should then take place within a year. The second IPCC assessment might therefore not be ready in time for the first conference of the parties and it became very important that the special report was completed in 1994. Discussions and negotiations in the INC of course continued in the meantime with priority given to defining more precisely possible agreements on implementation of the Convention, Article 21.2 specifying that

the head of the interim secretariat [of the Convention] referred to in paragraph 1 above will cooperate with the Intergovernmental Panel on Climate Change to ensure that the Panel can respond to the need for objective scientific and technical advice. Other relevant scientific bodies could also be consulted.

This reference to the IPCC was essential for further cooperation between IPCC and FCCC, which probably would not have been achieved, unless the IPCC had had the status of an intergovernmental body.

I considered it most important to seize on the possibility of developing further the spirit of collaboration that had characterised the joint work of the FCCC and IPCC so far. It seemed that a rather informal arrangement might best serve the purpose, at least to begin with. I proposed to the INC that a joint IPCC–INC/ FCCC working party be established, which was readily accepted by the INC. Ambassador Estrada-Ouyela and the INC executive secretary, Michael Cutajar, were invited to a first meeting in New York in November 1993. The cochairmen of the three IPCC working groups together with representatives of the key INC bodies were asked to attend this first informal consultation.

The INC chairman was naturally uneasy about the late completion of the SAR, since he had expected that a more precise interpretation of Article 2 of the Convention might have been possible on the basis of findings in the SAR. Further, a special meeting on this issue that had originally been scheduled for April 1994 in Fortaleza, Brazil, had had to be postponed until October that same year. The IPCC was unable to modify its schedule very much because of the heavy workload for the scientists engaged in the assessment processes, and the rules of procedures that had been agreed. Estrada-Ouyela expressed his worries in a talk to the Royal Geographical Society in UK:[5]

The scientists created the Convention . . . now when it is alive and walking, and deciding things, the scientists have reacted against its demands. The controversy reached the press under the heading "Frankenstein syndrome" hits climate treaty.

The exchange of letters between me and the INC chairman concerned whether the pace of the international negotiations was to be determined by the strict procedure needed for the acceptance of scientific assessment or by political factors. This disagreement was in fact just an indication of the growing importance of and urgency felt about the climate change issue and the IPCC's role in assessing what actually was happening.

To provide basic information for dealing with Article 2 of the Convention IPCC focused on developing stabilisation scenarios for greenhouse gas concentrations and on broadening this analysis to include socio-economic issues, thereby bringing another dimension to the scientific analysis.

The recognition that emissions of aerosols counteract the warming due to increasing amounts of greenhouse gases in the atmosphere lead to more uncertainty in determining the sensitivity of the climate system to human-induced forcing. This gave further ammunition to sceptics of the IPCC's work. Equity and fairness between people and countries also emerged as troublesome but of course important issues when socio-economic matters were put on the agenda, particularly because of the very large differences between rich and poor countries. This would necessarily have to be dealt with by the Convention.

7.3 Predictions or scenarios of future changes of the global climate?

Much of the discussion in the INC were thus focused on the question of how to achieve the prime objectives of the Convention as given in Article 2, i.e.

. . . to achieve stabilization of greenhouse gas concentrations in the atmosphere at a level that would prevent dangerous anthropogenic interference with the climate system. Such a level should be achieved within a time-frame sufficient to allow ecosystems to adapt naturally to climate change, to ensure that food production is not threatened and to enable economic development to proceed in a sustainable manner.

This in turn raised the issue of how sensitive the climate system might be to human interference and what actually can be said about future likely changes on the basis of scientific analyses. The general view amongst the scientists engaged in research in the field was rather pragmatic. Deduce with the aid of the best climate models available the changes of the global climate system when forced by past and prescribed future changes of greenhouse gas concentrations and aerosol loadings. Compare the results with the changes observed during the twentieth century. The agreement or disagreement between observed and computed changes would then provide a measure of the reliability of the model. In this way the best possible factual background for the political discussions in the INC would be provided. However, it was still difficult to tell how trustworthy projections of future changes might be. The key questions thus became: Are we really able to develop alternative credible scenarios for possible future human interferences with the climate system, especially as a result of future emissions of greenhouse gases into the atmosphere? Are we thereby able to outline the range of likely future changes of the global climate? These questions are legitimate. The climate system and the global socio-economic system are both chaotic and thus not really predictable very far into the future in the precise meaning of the word. My view of this complex issue at the time can be described as follows (see also Bolin (1997)).

The global climate system is to a considerable degree unpredictable although some gross features vary regularly. For example, the variations of incoming solar radiation bring about regular seasonal variations of the climate that can be simulated in their gross features with the aid of climate models. However, two experiments with identical initial conditions do not produce identical changes in the course of a year, nor are observed changes during two successive years identical. It is interesting to note that the observed *interannual variability of the global mean temperature* and that derived from model experiments that are only forced by the regular variations of solar radiation are about the same, i.e. about $\pm 0.2\,^{\circ}\mathrm{C}$, although the *variability of the local annual temperature* may be up to ten times larger. These are presumably the result of the random changes of the weather in the course of the year, spatial and temporal variations of the snow cover, extension of sea ice, periods of drought, etc. These natural variations are indeed unpredictable beyond about ten days.

Ice ages come and go and are largely triggered by changes in the distribution of solar radiation over the earth on a time scale from hundreds to the order of 100 000 years, but one ice age is never quite the same as another, partly because of slow but persistent changes of the land surface (for example glaciations) and the ocean circulation that may be induced. Only the gross features of this climate variability are understood today.

Similarly, seasonal variations are predictable with regard to their general features, but one winter is never the same as another one, especially so because the variations of weather are largely stochastic. However, weather statistics differ in winter and summer and thus depend on the changes of the gross features of the climate, i.e. seasonal variations, and can be approximately reproduced by climate models. One should therefore distinguish between the large-scale and usually rather slow trends of global climate change, for example, those due to external climate forcing on one hand, and on the other, the superimposed, irregular, largely stochastic, variation that will always occur and that partly hides the slow changes. This makes it difficult at first to detect a human-induced global climate change.

There are, however, also thresholds that may be passed, triggering more pronounced and rapid changes of the prevailing but slowly varying climate regimes (see Lorenz (1963)). This is known to have happened in the past. The ice sheets that covered the continents at high latitudes during the last ice age disappeared gradually between 12 000 and 7000 years ago, except for some temporary major setbacks. During the Younger Dryas period about 10 000 years ago, for example, a cold climate returned quite rapidly to the North Atlantic region and lasted for about 800 years. This change was probably caused by a decrease in the salinity of the North Atlantic surface waters because of the generally rather rapid melting of the ice sheets over Canada and Eurasia. This had a marked influence on the Gulf Stream, which temporarily returned to the more southerly position that it had occupied during the last ice age. In addition, it is interesting to note that all the ice sheets in the North Polar Region disappeared towards the end of the last ice age except the one that still covers Greenland. The precise reason for this is not well understood and we cannot predict if, and when, the ice sheets might disappear as a result of a future global warming, i.e. when a threshold of no return might be passed. The Greenland ice sheet now seems to be melting more quickly. If it does completely disappear, this may take centuries, but a gradual increase in sea level is expected. What gradual means is, however, still not well understood.

Nor are we able to predict changes in global society in the next century and thus how much our emissions of greenhouse gases might interfere with the global climate. To see this one merely needs to ask the question: how well would we have been able to predict the socio-economic development in the world during the twentieth century on the basis of what was known before the year 1900? We might have been able to foresee changes of some gross features for a limited period of time by extrapolating prevailing trends because of the inherent inertia of society. Similarly now, detailed features of societal changes during the next 100 years cannot be foreseen. In particular, human behaviour may change considerably. Human society is in this sense chaotic, particularly during periods of military conflict.

The interactions of the global human society and the natural environment and thus the climate system are even more complex and the use of the word *prediction* when communicating present knowledge about the climate change issue to the general public or politicians may indeed transmit a false impression of a capability that in reality is quite limited. These fundamental characteristics of chaotic systems are difficult for a non-specialist to comprehend. How then do we deal with Article 2 of the Convention?

We may be able to analyse the sensitivity of the climate system, i.e. determine its more immediate response, for a few decades, to specified external forcing such as changes of solar radiation or future emissions of greenhouse gases. One might thus determine the sensitivity of the system as dependent on the rate and magnitude of the forcing, but this would of course not be a prediction of climate change and its likely impacts. Nevertheless, this kind of partial information is of interest politically in order to have some idea about possible long-term trends of future changes, even if no details can be provided, except statistically. ᛁᛖ ᚬf ᚲᛚᛖ

The methodology should therefore rather be to present *sets of scenarios*. In these alternative assumptions about socio-economic development are made in order to deduce the forcing of the natural climate system that humankind might bring about and then let the non-linear behaviour of the combined natural and socio-economic systems play out.

This approach was already in use in 1992, when the six IS92 scenarios were deduced, although the behaviour of the societal system was at that time simply reduced to the specification of a few alternative scenarios for carbon dioxide emissions until 2100. The global climate models were also quite primitive. Feedback from the changes of the climate system on the socio-economic system and thus on possible changes of greenhouse gas emissions was not considered at all.

It should also be emphasised that it is not possible to base the development of socio-economic models on 'first principles', i.e. relations between key variables governed by the fundamental laws of physics, when developing climate models. Rather, statistical relations based on historical data have to be used and a number of assumptions about the future must also be made. These can be varied and a spectrum of 'possible futures' in the form of scenarios can be derived, but the range will of course become rather wide and expand as a function of time. How to interpret such a set of scenarios in a political context then of course becomes a crucial issue.

In the development of global socio-economic models the world is usually divided into a number of regions (in the case of the IS92 scenarios six were chosen, but later models have used more) that may be assumed to change differently. The initial division is, however, usually assumed to remain the same for the time span of the analysis, which of course will influence the long-term projections. For example, the UN projections of the future changes in the world population have usually been used. The economical development in the different regions has to be prescribed in terms of industrial development in general, future changes of gross domestic products, the rate of future improvements of energy end-use efficiency, the carbon intensity of the primary energy that is used, changes of land use and carbon emissions because of human exploitation of forests, etc. Obviously, none of the scenarios is a prediction, but the set as a whole provides a range of possible future emissions of carbon dioxide, as well as other greenhouse gases and aerosols, and therefore defines a likely range of the future changes of the climate system. However, as was so clearly expressed in the FAR, surprises cannot be excluded.

The IS92 scenarios that were presented in the IPCC 1992 Supplementary Report had in principle been deduced in this manner, but the approach was quite exploratory. Nevertheless, they provided a range of future path ways, and served as the basis for the SAR, since no better alternatives were readily available. Naturally, controversies arose about their interpretation and use. For this reason the IPCC decided to analyse them more closely and make the outcome available in time for the first conference of the parties to the Convention (see IPCC (1992d, 1995a)).

Because of the many assumptions that have to be made about the future development of global society, results from such analyses must be carefully scrutinised before being adopted as plausible possibilities. The attempts available when preparing the SAR were as yet few and modest in their aims. Economical considerations could be handled quantitatively more easily than social concerns in general, and the changing human responses to future developments and their consequences for society were largely ignored. The decision to develop a completely new set of scenarios was not taken until 1996 and they were actually not completed until the turn of the century. There were at the time considerable disagreements between the lead authors about the usefulness of the 1992 set of scenarios because of the limited attention paid to social matters. Nevertheless, because of their comparatively slow increase in the atmosphere, most of the greenhouse gases are globally well mixed, and it was felt that a limited number of scenarios spanning a reasonably wide range of global emission scenarios would certainly be of interest.

These scenarios were the subject of much discussion during the preparations of the chapter 'An evaluation of the IPCC IS92 emissions scenarios' (IPCC, 1995a), which was intended to supply information for discussions at the first meeting of the parties of the Climate Convention in March–April 1995. I had hoped for a more in-depth analysis than was finally produced, particularly with regard to key factors which determine the scatter of the scenarios that were available at the time. This was not possible because of the limited research efforts that had so far been launched and the assessment became a rather qualitative discussion of the scenarios available.

However, the key question had by then become: what limitations of greenhouse gas emissions would have to be imposed in order not to let the radiative forcing, and thus the climate, change to the extent that the objectives of the Convention as formulated in Article 2 would be violated? *Stabilisation scenarios* were required. Answers could easily be provided technically by so-called inverse calculations, even though the dynamics of the global carbon cycle were not very well understood, nor was it clear what socio-economic constraints would be required in order to achieve this in reality. It was later proposed that the costs incurred in order to bring about the required changes might be minimised (see Wigley *et al.* (1996)), but this had not yet been tried. Stabilisation scenarios were therefore derived in the special report 'Climate Change 1994,' that simply aimed at stabilising atmospheric carbon dioxide concentrations at 350, 450, 550, 650, 750 ppmv with the implicit understanding that the choice of a specific goal should be left for political negotiations.[6]

In the peer-review process of this assessment Donald Pearlman, and John Shlaes, leaders of the two key lobby organisations in the USA, requested major changes in the draft. Their criticism was based on disagreements about the purposes of emissions scenarios. They said that

... the terminology was ill defined, that the IPCC procedures were violated, and a grave departure had been made from the work schedule agreed in 1993 resulting in a lack of a comprehensive, fair, and balanced discussion of whether the IPCC 1992 scenarios are valid, appropriate and useful.

The view was also expressed that the IPCC did not have a mandate to refer to the possible use of the scenario analyses in the ongoing debate on sustainable development. Their intervention was a detailed legalistic examination of the purpose of and procedures for developing scenarios, which differed fundamentally from my own view, which was, as I understood, shared by many other scientists engaged in the assessment. The purpose of the exercise at this time could be no more than an effort to provide a broad idea about the long-term sensitivity of the climate system to rather widely different assumptions about future human activities. In addition, the climate scenarios presented in a useful way the sensitivity of the climate to external disturbances, the assumptions made and the range of total emissions that might be expected during future decades. Admittedly there were inadequacies in the draft that had to be corrected and terminology to be clarified, but the basic differences of views between the scientific community, on one hand, and representatives of many corporations and other stakeholders, on the other, about how to gradually improve our knowledge and understanding of the climate change issue were obvious.

The World Energy Council, led by its executive director Michael Jefferson from the UK, was another key international organisation that responded early to the potential threat of a human-induced climate change. Its attitude towards the issue was a much more constructive one. The World Energy Council initiated a special study that developed three scenarios for the world's future energy supply and used the associated carbon dioxide emissions as a basis for deducting human-induced climate change scenarios (World Energy Council, 1993). Initially only the next 25 years were considered. The work went beyond the IPCC efforts in that a more detailed analysis of the different technical options brought the climate issue better to the forefront amongst the leaders of the energy sector. The scenarios seemed more trustworthy, but the simple fact remained that the projections were still based on many assumptions that in one way or the other meant extrapolations of past changes. The projections were later extended to 2050 and 2100 in collaboration with the International Institute for Applied Systems Analysis (IIASA) in Vienna. The comments from the World Energy Council on the first draft of the chapter on scenario development in the 1994 IPCC special report were sharply critical and admittedly the World Energy Council scenarios were more informative. The IPCC scenarios were, however, still used by Working Group I in its work towards the SAR. My attention to the World Energy Council's efforts also later earned me an invitation to address their triennial world conference in Tokyo in 1995 (Bolin, 1996).

7.4 Attempting to put Article 2 of the Climate Convention into focus

It was obvious that the objective of the Climate Convention as expressed in Article 2 should be the basis for the development of a strategy for mitigation and scenarios were indispensable tools for the analyses of this issue. However, the workshop in Fortaleza in Brazil for in-depth discussions of these issues that had been scheduled for April 1994 had to be postponed for logistic reasons until October, by which time the Working Group III plenary session had already accepted the report on scenarios and approved the summary for policy makers for the special report. The timing of the IPCC work had not been satisfactory.

In my invitation to the Fortaleza workshop I had indicated that some additional workshops might be organised in 1995 with the objective in mind that these should be considered as preparations of a synthesis report that would be available for approval when the SAR was presented to the eleventh IPCC plenary session, then scheduled for December 1995. In the back of my mind was the very troublesome experience that I had at the fourth IPCC session in 1990 in Sweden, when I attempted to bring together the contributions from the three IPCC working groups into one single document. The preparations for the Fortaleza workshop were, however, squeezed between the finalisation of the special report on radiative forcing and scenario evaluation which was to be accepted and approved at the IPCC plenary session in November 1994, and the work of the writing teams preparing the drafts of the SAR. The papers to be presented in Fortaleza were not available in advance and some delegates complained that it had therefore not been possible to prepare adequately for the workshop. Again the conflict reflected the different approaches to a meeting of this kind by scientists and representatives of governments and government institutions. This Fortaleza meeting was to my mind to be viewed as a scoping meeting and additional meetings were going to be arranged during the first half of 1995 as part of the general preparations for the completion of the SAR.

Some 140 participants from 48 countries took part in the Fortaleza workshop. Many came from ministries and industry, which showed the political interest in this IPCC initiative. Even though the discussions were useful, the final outcome of the meeting was only a modest step forward. A number of general statements were made about what was needed, but the workshop did not turn out to be the best place to discuss how to deal with the political concept of '*dangerous anthropogenic interference*'. With reference to the lead authors of Working Group III, who had prepared the evaluation of the IS92 scenarios for the IPCC 1994 climate change report, I suggested in my statement to the conference that we should recognise two phases in the policy development (Bolin, 1994b):

1. *Short term (about two decades).* A precautionary principle should be adopted implying increased efficiency in energy production and use, reductions of emissions in cases when they cause multiple environmental damage, possible enhancement of sinks for greenhouse gases, research and development to provide new energy supply systems causing no or small emissions, and the development of adaptation strategies for minor but probably unavoidable changes of climate . . . It is, however, a political decision to decide how far-reaching efforts would be justified. A flexible policy should be adopted, whereby adjustments of temporarily agreed measures can be made.
2. *Long term.* As our knowledge increases, the basis for a long-term strategy will gradually emerge. Major efforts are desirable to permit scientific research to assist in an optimal way and to reach political agreements on long-term policies as soon as possible.

It was of course essential to have access to scenarios for this, but a new set would not be available until they were developed for the third IPCC assessment.

In retrospect it is not surprising that tangible results were few. It was premature to expect that more specific scientific analyses would be available to serve as a basis for a well-founded pursuit of the political issues as spelled out in Article 2 of the Convention, but a start was undoubtedly possible along the lines outlined above. In fact, 13 years later (i.e. in 2007) a short-term strategy is perhaps about to begin along the lines outlined above. No decisions with regard to a long-term strategy have as yet been taken, but these will be required in the negotiations regarding a second commitment period from 2013 and presumably extending to 2020 or even longer, but as yet nobody really knows. Climate politics develops very slowly.

7.5 Equity and social considerations

The negotiations that led to the adoption of a framework climate convention in 1992 were reasonably successful only because a clear initial distinction had been made between industrial (Annex-1) and developing (Non-Annex-1) countries foreseeing that their commitments for mitigation would have to be different for quite some time to come. The formulation of crosscutting issues as a key task for Working Group III implicitly also meant that the issues of equity gained a more prominent place on the IPCC agenda. This was also well reflected later in the introductory paragraphs of the chapter on equity in the SAR (IPCC, 1996c, Chapter 3):

Equity and social considerations are central to discussions of steps to be taken to implement the Framework Convention on Climate Change, both intrinsically and because widespread participation is essential if the objectives of the Convention are to be reached. Countries are unlikely to participate fully unless they perceive the arrangements to be equitable. This applies particularly to equity among regions and countries, but equity within countries and associated social considerations also influence what is possible and desirable. Mitigation and adaptation to climate change will require actions on the part of individuals. Governments will find it easier to comply with international obligations if their citizens feel that the obligations and benefits of compliance are distributed equitably. And richer countries are unlikely to burden their poorer citizens in order to benefit relative rich citizens in poor countries.

The overriding priority for the developing countries was of course to raise the standard of living of their populations and this would undoubtedly require more energy, which for quite some time to come would primarily have to be supplied by burning fossil fuels. This was also recognised in the Convention:[7]

The Parties should protect the climate system for the benefit of present and future generations of humankind, on the basis of equity and in accordance with their common but differentiated responsibilities and respective capabilities . . . The extent to which developing country Parties will effectively implement their commitments under the Convention will depend on the effective implementation by developed country Parties of their commitments under the Convention, related to financial resources and transfer of technology, and will take fully into account that economic and social development and poverty eradication are the first and overriding priorities of developing country Parties.

About 75% of the total *accumulated emissions* of carbon dioxide from 1860 to 1990 had come from industrialised countries and merely about 25% from developing countries, in spite of the fact that only a little more than 20% of the world population lived in industrialised countries in 1990. This meant that the contribution per capita to the accumulated emissions up to 1990 on average had been about 20-fold larger in industrialised countries. Their yearly emissions in 1990 were still about 64% of the global ones. These facts would obviously be powerful arguments for developing countries in the upcoming negotiations about restricting future emissions.

Of course, there are also other differences between industrialised and developing countries, such as affluence in general, the level of basic education and literacy, the resources for research and development, health care, etc. that justify a leading role for industrialised countries in limiting future emissions. It had already become clear during the preparation of the first IPCC assessment that controversies about these matters could become troublesome in the scientific assessments if the differences were not properly recognised and carefully evaluated. Efforts to assist developing countries to become involved in the IPCC work were obviously important.

Different views, however, also emerged amongst developing countries. There was naturally a tendency for delegates to focus on problems that were considered particularly important for their own countries. Small island states were concerned about sea level rise and pointed out that developed countries were the prime cause of this threat. Other tropical countries with forest resources were of the opinion that industrialised countries exaggerated deforestation in developing countries, implying that such activities were very significant in developing country contributions to the total emissions. This was not correct and the exploitation of the forests in equatorial countries, in particular in Brazil, Indonesia and tropical Africa had to be reduced. India, China and other countries in southeast Asia questioned the estimates of methane emissions from rice cultivation as summarised by Working Group I. Saudi Arabia, which led the oil-producing countries, consistently maintained that the whole issue of climate change was being exaggerated and this meant that well-founded scientific assessments were really still lacking (see Bolin (1993a)).

The IPCC was not yet able to tell very well whether some countries were more threatened than others. Climate models were not yet sufficiently well developed and tested to permit more firm conclusions of this kind. The issue could thus not then be resolved adequately beyond the obvious conclusion that poor countries were more vulnerable than rich countries and less able to protect themselves.

It was obviously important for equity also to be considered as a key issue within the IPCC itself. The IPCC Trust Fund, which had been created early on, took care of the costs for scientists from developing countries to participate in the assessment process, but a broader capacity building effort was also essential. The USA set aside $25 million in 1992 for studies of climate change impacts, adaptation and mitigation in developing countries, and financial support for this purpose was also provided by UNEP. In order to expand these activities the IPCC took the initiative in 1993 of asking for funds from the Global Environmental Facility (GEF), which had recently been established as a World Bank programme, in order to assist developing countries in their efforts to protect the environment. This initiative met some resistance, however, because the GEF's policy was to support specific research projects, while the IPCC's intention was rather to seek support for more substantial engagement by developing countries in the IPCC process.

Some support was finally received but the broader problem of capacity building in developing countries largely remained unresolved.

The participation by scientists from developing countries in the IPCC assessments certainly meant an increase in those countries' awareness and understanding of the fundamental scientific issues being addressed by the IPCC. I also sensed a positive response from developing countries in the IPCC Bureau with regard to the way their situation was appreciated by industrialised country participants. In particular, developing country politicians could acquire information directly from 'their own people' that was useful when participating in the INC and later also in the Climate Convention. The USA also supported countries in South America directly and the International Geosphere Biosphere Program (IGBP) made efforts to engage scientists from developing countries into the international Global Change research program that was being developed.

In order to emphasise further the importance of considering equity, amongst as well as within countries, IPCC invited delegates from developing countries in particular to a conference in Nairobi in July 1994 on the subject of 'Equity and Social Considerations'. The aim was primarily to underline that the IPCC's effort to provide basic scientific knowledge should recognise the importance of considering the major inequities between developed and developing countries in the further pursuit of assessing the climate change issue, especially with regard to impacts.

However, equity is a much broader issue than experienced in the course of the work by IPCC. Equity is essential when formulating the principles that govern the implementation of the agreements reached in the Climate Convention and the way countries fulfil their obligations in this context. It is obvious that the slow progress towards agreeing the Climate Convention during the first years of the twenty-first century is still related to the prevailing inequity in the world and the unwillingness of many industrialised countries to address that issue seriously.

7.6 Growing awareness of climate change and polarisation of opinions

The awareness of the climate change issue increased markedly, because of the agreement in 1992 to create an FCCC and the fact that 156 countries signed it at the UNCED in Rio, but most people did not bother to look at the detailed scientific analyses behind the conclusions drawn by experts in the field.

Sustainable development had been chosen as the key item for the agenda in Rio and it was expected that the attending countries would implement the agreed action. This was a very important change of the setting in which the IPCC pursued its second assessment. Industry largely responded negatively and questioned whether the available scientific knowledge was sufficient to justify major mitigation efforts at present aimed at slowing down and ultimately stopping the possible climate change. Economists maintained that the cost of mitigation would greatly exceed the benefits of avoiding damage because of climate change.[8] One principal problem is that the cost of the measures to be taken is expenditure now and in the more immediate future, while the benefits, i.e. avoided damage, will be experienced sometime in a more distant future and cannot be well quantified far in advance. They wanted to know how serious would a change of climate be, if no preventive action were taken. The outcome of the economic analyses also very much depends on the discount rate chosen for capital investments.

On the other hand, a number of environmental groups had expanded their activities to include the issue of climate change and asked for early action in order to avoid the catastrophe that they foresaw would seriously impact global society. The IPCC meetings were open to non-governmental organisations, and they were occasionally given the opportunity to express their views. A polarisation of the views on the climate change issue was clearly on its way.

When planning the UNCED in Rio in 1992 Maurice Strong, the executive secretary of the organising committee, initiated a closer look at the role of industry in sustainable development: this was led by Schmidt-Heine, an industrial leader from Switzerland. A number of case studies were presented, showing the advantage of being in the forefront of exploiting, in the positive sense of the word, the opportunities that emerging environmental concerns and issues of sustainable development might offer. Climate change could well qualify as a key issue in this context. It was viewed as a threat to industry globally, particularly to the security of the energy supply needed for increasing global industrialisation, and no obvious and immediate solution seemed to be at hand. The International Business Council for Sustainable Development was formed at about this time with the aim of bringing together like-minded representatives from industry with action-oriented activity in mind, but the responses were initially quite modest. As described in the previous chapter, I had met a rather sceptical attitude about the seriousness of the climate change issue at the congress of the International Gas Union in Berlin (Bolin, 1991) and the World Petroleum Congress in Buenos Aires (Bolin, 1992).

On the other hand, for rather obvious reasons the Dutch government had been quite active in the IPCC since the very beginning and Dr P. Vellinga, cochairman of subcommittee B of Working Group II, was in charge of much of the Dutch scientific involvement. During the first week of November 1993 a World Coast Conference was arranged in Amsterdam, focusing especially on measures to combat the rising sea level that might result from global warming. A few weeks later I was invited back to the Netherlands to give the Huygens Lecture,[9] an annual public lecture to the memory of the seventeenth century physicist and astronomer. It was interesting that the theme of global warming had been chosen and also that the podium for my presentation was the pulpit of a big church. This was undoubtedly an interesting experience, although I really wanted to deliver a scientific assessment rather than preach to a congregation! The proactive attitude created on these two occasions contrasted markedly with my experience in discussions with Dutch business people on a following day. They showed a striking reluctance to consider the issue of global climate change as a serious one, which of course was in no way a unique attitude, but nevertheless it was disappointing.

A number of individuals in the business community were also challenged by the IPCC assessments. Dr C. Starr, founding president of the Electric Power Research Institute (in the USA) criticised the carbon cycle modellers for their inference that the carbon dioxide uptake by the oceans is slow and said that since this was not true a rapid increase of atmospheric concentrations should not be expected in the future (Starr, 1993). He thought that the increase in atmospheric carbon dioxide concentrations so far was primarily due to natural variability. I found his analysis extraordinary. He mixed up the concept of the rate of carbon dioxide exchange across the air–sea interface, which is rather quick, and the gradual change of the equilibrium carbon dioxide concentration in ocean water because of the changing pH of sea water due to a net flux into the uppermost layer of the sea. In fact this feature was well known. I responded, since I had myself addressed this issue long ago (Bolin, 1995a, 1960) and it has been accepted by researchers in the field ever since. What was the intention of this manipulation of well-known scientific findings?

A similar attempt to discredit the IPCC analyses was published by Dr H. Linden, Illinois Institute of Technology, executive advisor of the Gas Research Institute (Linden, 1993). He enumerated in his article selected features of the global climate system, but did not attempt to put them together into a more organised and trustworthy picture. He did not accept that human emissions had caused the increase of atmospheric carbon dioxide concentrations (and in doing so referred to Starr's report), he rejected the careful analysis of the changes of the radiative fluxes that an increase of greenhouse gas concentrations would bring about, he referred to local and global changes of temperature in a most confusing manner and he accused the scientists behind the IPCC reports of painting pictures of catastrophe, but did not explicitly give a single example of what he meant IPCC had said in this regard. In fact, it was the media that painted scenes of catastrophe, while the IPCC reports had been carefully vetted to avoid this kind of criticism.

I was greatly troubled that key individuals in industry rejected careful analyses by scientists in the field using scientific arguments that could so easily be proven to be false, but that this might escape detection by the non-specialist. There were many more opponents than the few mentioned here. This contributed to the polarisation of public opinion that was at that time not so obvious but that grew during the coming years, particularly after the publication of the IPCC SAR in 1996, to which I will return later.

The assessment of the progress in the socio-economic sciences regarding climate change was the responsibility of Working Group III and a key to the work was to involve classical economists as well as researchers who saw environmental economics as a new field of research and recognised the environment as a resource for the sustainable development of humankind that needs special protection. It was also important that the legitimate wishes of developing countries to eradicate poverty and truly aim for equity amongst people were kept in focus, even if efforts to combat climate change were placed high on the agenda. By 1994 the work of the IPCC was widely recognised and there was an increasing interest in analysing its mode of operation, which had brought the global issue of climate change to the forefront of UN activities in just five years.

In November 1994, just as the tenth IPCC session (IPCC, 1994b) was being held in Nairobi, Kenya, Sonja Böhmer-Christiansen at the University of Sussex, UK, published her view of the emergence of the climate change issue, in which a fight for influence and power between the scientific community and politicians is described (Boehmer-Christiansen, 1994a,b). Certainly, there are elements in her analysis that I recognise, having been in the midst of this development for a good number of years, but she drew a number of conclusions that were based merely on superficial observations and she obviously lacked knowledge about what actually had happened. In addition her writing had an insinuating tone that seems to be based on preconceived ideas of the issues being discussed.

The possible threat of a global climate change and what it might imply for people and countries was hardly touched upon. After all, the prospect of a major climate change would have to be seriously considered in itself and in the context of sustainable development. The discussion of the uncertainties of the assessments and the efforts made by the IPCC writing teams to formulate the scientific conclusions reached in the scientific literature in as accurate a way as possible was not accepted, nor was the fact that ongoing research might well make some earlier conclusions invalid and these might therefore be reconsidered in a later assessment report, when such new findings have been published and scrutinised by the scientific community. The assessment process is in this sense necessarily a dynamic one. This basic fact was missed in her analysis.

She considered the personal acquisition of increased financial resources for research as the prime driving force for the interest shown by the individual scientists, not their engagement in providing new knowledge in a central research field that might be of societal significance. It is, of course, essential to secure financial support for scientific research and there is competition, but I attended many writing team meetings and was struck by the genuine engagement in the scientific issues being dealt with that was shown. Lack of resources was hardly an issue when the best scientists met to contribute to the task pursued by the IPCC. It is just impossible that coordinated efforts to emphasise uncertainty were in some way agreed upon in order to acquire further funding.

Her misconception about the role of the two parent organisations, UNEP and WMO, was also striking. In reality they provided a neutral platform for the interplay between scientists, government representatives and politicians at the IPCC plenary sessions, while the outcome of the assessment process was largely the result of the involvement of some of the very best scientists from their respective fields. This could only be achieved by ensuring impartial and high quality efforts. UNEP and WMO had themselves minimal resources for research and no influence on the outcome of the assessments. They had created the IPCC, but the IPCC was itself an intergovernmental scientific body that could largely act independently in its task of assessing the available knowledge regarding the global climate and its variation through the work of the assessment teams. The IPCC working group bureaux had an independent status, for example when choosing lead authors for the reports, which is an important feature of the structure that was in existence from the very beginning. Planning of cooperative global research efforts, on the other hand, was, and still is, the responsibility for the WCRP and the IGBP in cooperation with the non-governmental International Council of Science, ICSU.

Boehmer-Christiansen also expressed the view that politicians had made themselves too dependent on scientific advice, expecting that the uncertainties would be resolved if research was intensified. Further research will provide an improved basis for decisions, but knowledge will always be incomplete. It is therefore important that as far as possible reasonably robust conclusions are extracted. Above all, dealing politically with a complex issue like climate change without a thorough analysis of what we do and do not know would be a disaster.

Finally, the concluding paragraph of her article was an accusation that the independence of the IPCC had not been safeguarded, that politicians were not fully aware of the complexity of the climate issue and that the international process that has emerged is not trustworthy, in spite of the fact that the status of IPCC was widely appreciated, not least by the scientific community at large.

Under pressure, even scientists will deliver what their paymasters prefer to hear. Honest science may be less useful in the short term, than relevant science, a lesson that has to be continuously relearned. Policy related advisory networks need to become more sophisticated and less self-serving – and policymakers to develop broader decision-making structures – to prevent this from happening. This cannot mean listening to green or energy lobbies to the exclusion of all others. Advisors and decision-makers need each other, but knowledge funded by soft money and created under conditions of dictated relevance and competitive bidding is surely unlikely to inspire the degree of trust upon which wise policy must be based.

Political decisions and actions have to be based on value judgements of *all available relevant information*. This often implies a balance between opportunities and fears, but this is not a task for the scientific community. I expressed my thinking in this regard at an IGBP conference in Ensenada, Mexico, in 1992 (Bolin, 1994a). Politics must not be paralysed by uncertainty, which is part of life. We must rather learn to live with uncertainty. This same view was later well expressed on the first page of the summary for policy makers for the SAR from Working Group III:

Analyses indicate that a prudent way to deal with climate change is through a portfolio of actions aimed at mitigation, adaptation, and improvement of knowledge. The appropriate portfolio will differ for each country. The challenge is not to find the best policy today for the next 100 years, but to select a prudent strategy and to adjust it over time in the light of new information.

The best possible analysis of scientific information is essential in order to do this well.

7.7 The approval of the 1994 IPCC special report runs into difficulties

Working Groups I and III accepted the special report on radiative forcing and climate change and the evaluation of the IPCC IS92 emission scenarios, and approved the summaries for policy makers during September and October 1994. Considerable difficulties were encountered, however (IPCC, 1994a,b).

The draft chapters of the report by Working Group I had gone through the scientific peer-review process and note had also been taken of government comments when revising the draft, but the final proposed version was not generally available at the Working Group I session in Maastricht when the summary for policy makers was to be considered for approval. This was formally a breach of the rules of procedures. This was, however, taken care of at the session. The meeting's delegates agreed unanimously to a procedure by which unresolved issues would be considered at the IPCC plenary meeting a few months later if so required. In addition, however, the IPCC press release was incorrect and was actually released before the meeting had come to an end. It had not been approved properly by the cochairmen of the working group. They apologised for the mistakes, but it still caused unnecessary misunderstandings, which was most unfortunate.

However, the way these incidents were used in order to discredit the IPCC vis-à-vis the US Administration and the US Congress was most troublesome. The executive directors of the Global Climate Coalition and the Climate Council, John Shlaes and Don Pearlman respectively, wrote a long letter, supported by the presidents and vice-presidents of six of the leading private industrial organisations, to the US delegates at the working group session[10]. I quote:

It is apparent that those responsible for issuing the press release were trying to use the IPCC to persuade policymakers and the public of their personal views concerning climate change scientific and policy issues, even though that view does not represent a consensus . . . We believe that the United States, which provides a disproportionate share of the funding for the IPCC . . . has the responsibility to exercise leadership to correct this unfortunate situation and to ensure that it is not repeated in the future . . . We have participated extensively in its [IPCC's] activities through several of our organizations and hope that, with IPCC adherence to fair and proper procedures, we can continue our efforts.

The cochairmen to the working group regretted what had happened and the working group agreed that the rules of procedure would have to be followed more closely. However, no delegation considered the issue serious enough to take any further measures.

Soon thereafter five leading, sceptical scientists wrote a letter to the Undersecretary of the US State Department, T. Wirth, expressing their disappointment in the press release issued from the Maastricht meeting.[11] They even accused the chairman of Working Group II, Watson, of not defending the interests of the USA, which is a strange accusations to make of a scientific analysis, even in the case of a summary written for policy makers. In a written response, I, together with the two cochairmen of Working Group I, expressed our regrets concerning the inadequate preparations of the press release, but we rejected the accusations that the IPCC reports had been partial and that they should not be considered as representative of the views of the scientific community. On the contrary, dissenting views had been recorded and uncertainties dealt with carefully and adequately. However, the Maastricht incident was a reminder that procedural issues must not be ignored. I am sure, however, that there were no significant inaccuracies in the report when it was unanimously accepted, nor in the summary for policy makers that was formally approved. It should also be noted that the Undersecretary of the US State Department T. Wirth did not accept the accusations as valid and political judgements had obviously been made high up in the Administration on this issue. Al Gore served as Vice-President from early 1993. His engagement in the climate change issue had become obvious from his book, *Earth in Balance*, published the year before (Gore, 1992). He was certainly kept well informed at this time by the scientific advisor in the White House, J. Gibson, and his assistant, Robert Watson.

Working Group II ran into trouble about the way that it should deal with the impacts of climate change. The lead authors naturally maintained the view that positive as well as negative consequences should be described, but this met with resistance from several developing countries who cited the way this matter was dealt with in the text of the Convention. The Convention consistently stressed only the need to consider the *adverse effects of climate change*. Obviously this biased view was not acceptable and was not agreed by the working group.

In the plenary session of Working Group III (in September 1994) the IPCC scenarios were again attacked by sceptics because there was no way to judge their quality. This led to heated discussions between delegates from oil-producing countries (supported by lobbyists from the USA) and representatives from some other countries. The session ended in confusion and it was decided that discussions should be resumed in a session immediately before the IPCC plenary session in November. At that time I imposed stricter rules for the debate and the summary for policy makers was finally unanimously approved.

Even though a number of difficulties were encountered in the pursuit of the IPCC assessment and there were increasing doubts in industrial circles about the seriousness of the climate change issue, the ratification of the Climate Convention had progressed more quickly than expected. The fact that it had come into force early in 1994 meant there were new bases for the negotiations, but agreements on joint measures turned out to be very much more difficult to reach.

7.8 Preparing for the future role of the IPCC

In view of the fact that the first conference of the parties of the Climate Convention was to be held in March–April 1995 I considered it very important that the structure of the IPCC and its modes of operation should be carefully reconsidered. The Convention was going to create two subgroups, the Subsidiary Body for Scientific and Technical Advice, SBSTA, and the Subsidiary Body for Implementation, SBI. It was important to clarify the future role of the IPCC in relation to these two new bodies and particularly the first one. I realised again at that time the importance of the IPCC being an intergovernmental body, that was nevertheless able to act independently of its supporting countries and their national interests, i.e. to assess available knowledge in a trustworthy manner. After discussions I prepared a discussion paper for the IPCC tenth session (IPCC, 1994b). The following key recommendations were unanimously agreed:

The IPCC should be retained as the prime independent body to provide the Conference of the Parties with up-to-date scientific, technical and factual socio-economic assessments of current relevant knowledge.

Because of the scientific and technical nature of the assessment – and hence clearly separated from policymaking – it is proposed that the IPCC remains as a jointly sponsored body of UNEP and WMO, and that its role as a prime body for assessing scientific and technical issues for the FCCC be agreed by the Parties to the Convention.

The next full assessment by the IPCC of the climate change issue should be completed by the year 2000 and detailed plans and schedule be decided in 1998.

The IPCC should assess particular topics in response to the needs of the FCCC and requests from the sponsoring bodies, and in response to scientific or technical questions arising in the literature (it is envisaged that the Panel would meet at intervals of about one year . . .).

The IPCC activities regarding methodology development should as of the end of 1995 be restricted to periodic assessments of ongoing work in order to ascertain global consistency based on knowledge available in the scientific literature. Development of new methodologies should be undertaken as the need for them arises.

This implied that the responsibility for developing the necessary arrangements with national agencies for the purposes of these tasks would be taken over by the Convention secretariat.

8

The IPCC second assessment report

The polarisation of the views on the reliability and adequacy of the scientific and technical knowledge base increase.

8.1 First party conference of the FCCC

It is important first of all to make clear that the Climate Convention that came into force in 1994 is a framework convention that primarily specifies the procedures to follow and agreements on the general structure of the intergovernmental arrangements required in order to deal with the climate change issue. It does not specify any quantitative and binding commitments for the parties, but regulates the important matter of establishing a reporting system between countries and the Convention secretariat.[1] Two prime tasks for the first conference of the parties in Berlin in March–April 1995 were accordingly to elect chairpersons and members of committees to be set up, and to formulate the goals for the negotiations between countries during the next few years, i.e. the 'Berlin mandate'. As chairman of the IPCC I was also anxious to get a clear idea about the forthcoming interplay between the IPCC and the SBSTA which was to be formed by the Convention.

The upcoming first conference of the parties stimulated discussion in wider circles of the assessments carried out so far by the IPCC, and the process for the assessment that IPCC had developed. There would obviously be more publicity about the climate change issue and also the IPCC during coming months. I was going to give a presentation at the UN at the beginning of February and it was decided that I would then spend a few days in Washington DC, and thereby become more acquainted with representatives of the US press. It was also planned that I would meet with some members of the US Congress. The arrangements were taken care of by staff at the World Resources Institute.[2] This gave an

106

interesting (although certainly incomplete) insight into the way the climate change issue was handled in the USA.

It was obvious that views amongst journalists were often based on a rather superficial acquaintance with the work by the IPCC. It is for example noteworthy that even the renowned journal *Nature* misinterpreted the IPCC's mode of operation and several of the key conclusions published in its reports. An editorial just a few weeks before the conference of the parties began contained a number of misunderstandings.[3] For example, it was maintained that the work of the IPCC reflects 'the vested interests of its sponsors', presumably referring to the UN organisations WMO and UNEP. This view might stem from the article by Sonja Böhmer-Christiansen, published in *Nature* a few months earlier (see Chapter 7.6). In a letter to *Nature* the cochairman of the Working Group I, Sir John Houghton, and I responded (Bolin and Houghton, 1995) that 'the IPCC determines its programme of work and publishes its assessments with no interference from the WMO and the UNEP', which merely pay for an administrative secretariat in Geneva, which also keeps the IPCC accounts straight.

Another presumed 'weakness' in the IPCC process was also stressed in the editorial '. . . that the IPCC materials are not comprehensive nor welcoming contrary opinions.' We responded:

Anyone who reads our full reports cannot fail to be impressed by their comprehensiveness. Regarding contrary opinions, the review process which is undertaken by IPCC specifically addresses conflicting scientific opinions and ensures that these are properly exposed. The particular example quoted in your leader concerns the role of water vapour which, because of its importance as a greenhouse gas, is crucial in any assessment of global warming. It is stated that water vapour (especially upper level water vapour and ice) is neglected in climate models. This is incorrect as is evidenced *in all of the IPCC reports* [emphasis given here]; models which have been available during the last five years include comprehensive description of all phases of water and their role in cloud formation. Further, because of some significant controversy over the magnitude and sign of water vapour feed back (associated with the 'air conditioner' idea mentioned in your leader), the IPCC 1992 report discussed the role of upper tropospheric water vapour in some detail.

This misconception about the role of water vapour was actually quite wide spread at the time, but it is remarkable that this issue was brought up by *Nature*, obviously without checking in the reports what the IPCC had actually done. Lack of insight of what an assessment of the issue of global climate change really amounts to was also reflected in the challenge of the IPCC role with the proposal that

. . . the United Nations has a better model to follow, that of the UN Scientific Committee of the Effects of Atomic Radiation (UNSCER), which has over several years produced a series of technical assessments that carry general conviction . . . It [the steering committee that would be required to deal with the climate change issue] should be content with assessments written in technical language; governments have plenty of technical people to advise them of the importance of what emerges.

The global warming issue could not possibly be dealt with by a small technical committee in the manner envisaged. The broad scientific community concerned with the science of climate change, adaptation to it and mitigation to keep it within bounds, necessarily demands a much more participatory setup. It had also become very clear that openness towards stakeholders of different kinds, not least the environmental organisations, and the public might well become an absolute necessity, and the assessment had to be carried out in a scientifically independent manner. On the other hand, some very pertinent comments on the role of developing countries in the attempts to deal with a human-induced climate change were published by *Nature* after the Berlin conference.

Expectations of what it might be possible to achieve at the Berlin conference grew, but many realised that it was unlikely that much more than the formalities would be dealt with on this occasion. At the conference rhetoric was common place and the tension between developed and developing countries increased. For example the Alliance of Small Island States (AOSIS) proposed explicitly that developed country emissions should be reduced by 20% by 2005, which was a completely unrealistic demand at the time, but it expressed of course the sense of anxiety and urgency that was felt. In a letter to the conference India expressed its great concern that so little had been done by developed countries since the agreement on a Convention had been reached in Rio. A similar sentiment was put to the conference by the Chinese delegate. I was invited to a meeting of environmental organisations on the day before the conference opened to inform them about the IPCC's work on its second assessment, which I of course welcomed. On this occasion the German minister responsible for environmental issues, Angela Merkel, also spoke. She was forward looking but referred to the probable increasing frequency of extreme events which had already become more common, although evidence for this was very meagre.

Chancellor Helmut Kohl opened the conference and, Angela Merkel, in her position as the chairperson of the meeting, stressed in her opening statement the perhaps most important aspect of the climate change issue at the time, namely:

In accordance with the principles of the Convention, particularly in view of the common but differentiated responsibility, we, the industrialised countries, must be the first to prove that we are bearing our responsibility in protecting the global climate. Only when we demonstrate this by convincingly taking the lead, can we demand actions from other countries regarding climate protection. After all, we are talking about preserving our *single* world, as we repeatedly stressed in Rio.

As chairman of the IPCC I had been asked by the chairman of the INC to address the conference. After the approval of the special report, *Climate Change 1994*, at the IPCC tenth plenary session (IPCC, 1994b), my attention had been directed towards the IPCC making a more specific contribution to this first conference of the parties, even though the IPCC would not, and should not, be directly involved in the negotiations.

The IPCC conclusions in 1992, on which the Convention was based, now seemed perhaps even more convincing to many delegations than when adopted, at least it seemed so from the speeches delivered. I therefore wanted to remind the delegates of the wording in the Convention (Bolin, 1995b):

Parties should take precautionary measures to anticipate, prevent or minimize the causes of climate change and mitigate its adverse effects. Where there are threats of serious or irreversible damage, lack of full scientific certainty should not be used as a reason for postponing such measures . . . but it is of course also important to analyse the possible implications of long-term preventive actions and in doing so recognise [that] economic development [should] proceed in a sustainable manner.

I felt that these principles were very important, but regretfully, though perhaps not surprisingly, they had still only partially been dealt with in 2007, more than ten years later. I also stressed the rapidity with which greenhouse gas concentrations were actually increasing:

. . . about 70% of the increase of the atmospheric carbon dioxide concentration [so far] is due to the emissions that have occurred during the last 50 years. The IPCC basic scenarios of future emissions show that the same amount will be emitted again [but now merely] within about 25 years, if no preventive measures will be taken.

The rate of change is still (in 2007) accelerating, although less so, but there is a long way to go before a decrease of emissions is in sight. It may also be interesting to recall my assessment that

. . . the magnitude of the human interference with the climate system becomes clear when realising that the global [enhanced radiative] forcing due to human greenhouse gas emissions presently corresponds to about 1% of the solar energy that is absorbed by the earth. This amount of energy is about one hundred times larger than today's [human] use of energy in the world as a whole.

This indeed brings home the efficiency of the enhanced greenhouse effect. My conclusions and challenges to the conference were:

The climate change is a long-term issue. The global climate system responds to anthropogenic greenhouse gas emissions with a delay that may be several decades, which means that early detection of serious threats is difficult. Also, when changes are on the way, major reductions in forcing factors may well be required in order then to change course.

The socio-economic system, however, can only be changed gradually in order to give societies time to adjust and accept changes. And above all, while initial measures may not involve large costs, late short-term interventions, if [later] required, may be much more costly. The need for possible early actions should be carefully assessed, keeping in mind that they must not seriously compromise necessary development. The responsibility for that judgement rests with you as representatives of the governments that have agreed to the spirit and letter of the Convention. I can assure you that the IPCC scientific/expert community will be following your endeavours closely and remain ready to assist.

The IPCC was in the midst of its efforts to complete the SAR and its input to the conference had therefore been limited to the special report, *Climate Change 1994*, that was approved in November 1994 (IPCC, 1995a). Few national delegates referred, however, to this or other IPCC reports in their statements at the plenary sessions. Politics dominated the discussions. Steps beyond what had been agreed in the Convention were not to be expected. The focus was rather on the formulation of the Berlin Mandate, i.e. the basis for negotiations to be conducted in order to prepare for the third conference of the parties that was scheduled for Kyoto in December 1997. It was unclear, however, how far-reaching a goal the conference might agree on as a first step towards action.

Developing countries, acting jointly under the heading G77, were obviously not willing to take on compulsory commitment in a first round of action. A controversy was developing because of the USA's reluctance to accept such a generous attitude by industrialised countries. The head delegate from USA, Tim Wirth, was squeezed between conservative members of the US Senate, who attended the conference as observers, and the delegates from other countries that had primarily listened to the environmentalists representing non-governmental organisations.

The negotiations were skilfully led by Angela Merkel, who succeeded in reaching a unanimous agreement that no quantitative targets would be imposed on the G77 countries until the first commitment period 2008–2012 had come to an end. The industrialised countries should rather be in the forefront and accept a first set of preventive measures, while developing countries in the mean time would be given an opportunity to develop on a course towards sustainable development.

Reluctance to accept quantitative obligations had also been expressed at an internal EU meeting by some member countries, primarily from the Mediterranean region, but most delegates realised that a collapse of the negotiations would be seen as a failure for Europe, which had been in the front line defending the aims of the convention. The UK was in the EU chair at the time and the minister responsible for environmental issues, Mr John Gummer, spoke out very strongly and effectively countered this potential split amongst the European countries. This was essential because most other industrialised countries were hesitant and the conference could well have ended without a reasonably clear and ambitious goal for the next 2½ years having been agreed. It might also have been seen as a failure for the IPCC not to have been able to emphasise adequately that both the global climate system and the socio-economic structure of the world are systems with great inertia that needs to be recognised. Action was indeed urgently needed.

8.2 The IPCC Second Assessment Report

The work on the IPCC SAR by the three working groups was somewhat delayed because of the efforts to complete the special report in 1994, particularly so for Working Group I. The difficulties experienced in the process of approving its report in 1994 signalled that controversies might well arise again. It was clear that US lobby groups, particularly the Climate Council and the Global Climate Coalition were preparing for the next IPCC session at the end of the year. In spite of the difficulties that were experienced when the IPCC FAR was considered for approval, I still considered it important to aim for a synthesis report to be considered at the plenary session in December.

Major efforts were required by the scientists engaged by the three working groups to finalise the voluminous bulk reports (they ultimately comprised some 2000 pages and contained about 10 000 references to the scientific literature). I will necessarily be selective when summarising the many discussions that went on, because of the vast quantity of literature covered. Particular attention will be given to issues that later became controversial and that figured prominently in the public debate during 1996–7. For a more detailed technical account the reader is referred to the three volumes of the IPCC SAR, and particularly the summaries for policy makers, prepared by the working groups, as well as the synthesis volume (see IPCC (1996a,b,c,d)).

8.2.1 *Working Group I assessment report,* Science of Climate Change: *approved in Madrid, 27–29 November 1995*

The key findings of the Working Group I report on the science of climate change concern the overarching issues of detection of climate change, the determination of its characteristic features and the reasons for its occurrence. The following conclusions are particularly noteworthy:

- Atmospheric greenhouse gas concentrations continue to increase, leading to an increasing positive radiative forcing of the climate system. The concentration of carbon dioxide was 358 ppm in 1990, an increase of about 28% since preindustrial times. The temporary slow-down of the rate of increase in 1990–3 (see IPCC (1995a)), partly due to the Pinatubo volcanic eruption, has come to an end.

- The direct radiative forcing of the long-lived greenhouse gases was 2.45 W m^{-2}, of which about 65% was caused by the enhanced carbon dioxide concentration.
- The enhanced concentrations of troposphere aerosols resulting from combustion of fossil fuels, biomass burning and other sources caused some negative forcing, globally about 0.5 W m^{-2} directly, and in addition through the change of the cloud albedo (reflectivity), although its global magnitude is poorly known. This negative forcing can be *locally* large enough to more than offset the positive forcing due to greenhouse gases.
- The global mean temperature has increased between 0.3 and 0.6 °C since the late nineteenth century.
- Regional changes are also evident. For example, the recent warming was greatest over the mid-latitude continents in winter and spring, with a few areas of cooling, as over the North Atlantic Ocean.
- The global sea level has increased between 10 and 25 cm over the past 100 years and much of the rise may be related to the increase of the mean temperature of the oceans.
- There is inadequate data to determine whether consistent global changes in climate variability or weather extremes occurred during the twentieth century.

The last conclusion should be noted because the public view of the global warming issue was primarily based on reports about enhancements of climate variability and more frequent occurrences of extreme events. The sceptics, who did not accept the reality of an ongoing human-induced climate change, on the other hand, believed that the IPCC assessments exaggerated the risks of a human-induced climate change, which scared the public. They provided, however, few if any references to specific parts of the report that would support such views. However, it was non-governmental groups of environmentalists, supported by the mass media who were the ones exaggerating the conclusions that had been carefully formulated by the IPCC.

The following conclusions were also agreed but only after a heated debate:

Any human-induced effect on climate will be superimposed on the background 'noise' of natural climate variability, which results both from internal fluctuations and from external causes such as solar variability or volcanic eruptions . . .

Our ability to quantify the human influence on global climate is currently limited because the expected signal is still emerging from the noise of natural variability, and because there are uncertainties in key factors. These include the magnitude of patterns of long-term natural variability and the time-evolving pattern of forcing by, and response to, changes in concentrations of greenhouse gases and aerosols, and land surfaces changes. *Nevertheless, the balance of evidence suggests that there is a discernible human influence on global climate* [emphasised here].

This second paragraph quoted led to discussions that lasted a day and a half. One of the convening lead authors of the Chapter 8 of the report (Ben Santer) opened the discussion by presenting new evidence that would justify a stronger statement regarding a partial attribution of the observed change of the global climate being due to human interference than had been proposed in the wording submitted in writing before the meeting of the working group, and he was supported by the other convening lead author (Tom Wigley).[4] The very last sentence in a first version, however, contained the expression 'appreciable human influence' rather than 'discernible human influence.' I felt uneasy at this as did the UK delegate (David Warrilow). After further consultations, I proposed to the chairman that the discussion be reopened and that the wording of the last sentence of the crucial paragraph be modified by using the word 'discernible' rather than 'appreciable' as had first been agreed. There were no objections to this proposal which better emphasised the uncertainty. Even though the precise meaning of the word 'discernible' was still somewhat unclear, to my mind it expressed considerable uncertainty as well as the common view that it was impossible to provide a more precise measure. The modification also satisfied Dr Al-Sabban (Saudi Arabia), who had objected to the earlier wording. He did not, however, consider whether the proposed conclusion was adequately supported in the underlying bulk report available at the meeting. The chairman, Sir John Houghton, pointed out that since the two convening lead authors considered that this modification was justified in the light of the additional literature that had been considered by them[5], a modification of the writing in the bulk report and the insertion of additional references were needed in order to ensure consistency. This was unanimously agreed by the working group. The modifications of the bulk report that were accordingly made after the meeting were agreed by the team of lead authors of the chapter as well as by the working group chairmen before final publication. Most importantly, the final wording of the last sentence was in my view well chosen on the basis of the evidence that was available at the time.

This was indeed an awkward interlude because the written submissions to the session were supplemented during the final session, but in retrospect it was important in order to ensure that the work by the scientists would not become unduly constrained and bureaucratic. It should be emphasised that there was no expression of mistrust in the scientific conclusions that were drawn by the convening lead authors. For most delegates the issue was one of nuances of expression, while there were just a few objections raised about the procedure, though none by the US delegation.

The following additional conclusions of the working group regarding projections of future climate changes are also pertinent:

For the mid-range IPCC scenario, 1992a, assuming the best estimated value of current climate sensitivity and including the effects of future increase of aerosols, models project an increase in global mean surface temperature relative to 1990 by about 2 °C by 2100. This estimate is about one-third lower than the 'best estimate' in 1990 [i.e. in the FAR].

This is due primarily to lower emissions scenarios (particularly for CO_2 and the CFCs), the inclusion of the cooling effect of sulphur aerosols, and improvements of the carbon cycle. [A range of uncertainty, 1.0–3.5 °C, was also provided] . . . Confidence is higher in the hemispheric-to-continental scale projections of coupled atmosphere–ocean climate models than in the regional projections, where confidence remains low. There is more confidence in temperature projections than hydrological changes.

All model simulations . . . show the following features: greater surface warming of the land than of the sea in winter, a maximum surface warming in high northern latitudes in winter, little surface warming over the Arctic in summer; an enhanced global mean hydrological cycle, and increased precipitation and soil moisture in high latitudes in winter. All these changes are associated with identifiable physical mechanisms. [The Working Group also gave the warning that] many factors currently still limit our ability to detect and project future climate changes.

Although the discussions were intense during the final session of Working Group I, and in spite of the fact that some parts of the summary for policy makers that had been presented to the working group for consideration at its final session had to be deleted and rather included in the technical summary because of insufficient time, an important step forward was taken in the attempt to reach agreement on key features of the ongoing climate change.

The summary for policymakers was approved and the technical summary and the bulk reports accepted and this was all brought to the IPCC plenary meeting in Rome in December for final acceptance. Nevertheless, a confrontation had occurred between the scientists who wished to put forward their latest findings as soon as possible and some of the delegates who emphasised more the need to safe guard the credibility of the assessment process. This issue resurfaced in the course of the following years.

8.2.2 *Working Group II assessment report,* Impacts, Adaptation and Mitigation: *approved in Montreal, 16–20 October, 1995*

Working Group II's assessment was concerned with trying to summarise know-ledge about the *impacts* of the global climate change that might result in the alternative emission scenarios that were developed, and the possible needs and means for adaptation and mitigation. The analyses of systems aspects of these issues were primarily dealt with by Working Group III.

Few attempts had yet been made to deduce more detailed patterns of climate change that might result from projected changes of greenhouse gas concentrations. There were therefore complaints from scientists engaged in impact analysis projects that it was simply not possible to go much beyond the determination of the sensitivity and vulnerability of various regions in the world to different specified changes of temperature and precipitation, which were viewed as the most crucial factors, particularly for agriculture and forestry. However, some attempts were also made to provide regional and even global views of the impacts. Possible changes in the distribution of ecosystems were derived for the case of a climate change resulting from doubling of the atmospheric carbon dioxide concentration.[6] Although such maps were of great general interest, they could not in practice easily be made use of, since local information (for example, the distribution of soils) is required to interpret them and this can only be supplied through more detailed national studies. Few countries had initiated such work and the information available did not easily translate into a general assessment of the local and regional implications of a gradual change of the ecosystem distribution.

The assessment of possible means of mitigation was another of Working Group II's tasks and considerable interest was shown in the analyses of options to reduce emissions and enhance sinks of greenhouse gases. Not surprisingly, this caught the interest of the industrial sector, and once again controversies arose. The main reason was the attempt by the lead author team (with 24 members[7]) to describe scenarios of decreasing carbon dioxide emissions by projecting the gradual introduction of new technology during the twenty-first century.

To assess the potential impact of combinations of individual measures at the energy system level in contrast to the level of individual technologies, variants of a Low CO_2 Emitting Energy Supply System (LESS) were described. The variants of the LESS constructions should be viewed as thought experiments exploring possible future global energy systems.

A number of assumptions had to be made about changes of the global population, expected increases of the gross domestic product during the twenty-first century and the differences between developed and developing countries in this regard, possible rates of decarbonisation of the energy system and the introduction of various forms of renewable energy, above all biomass. A low energy demand variant (as given in the IPCC 1990 assessment) was chosen and emphasis was also given to energy efficiency (i.e. primary energy consumption might rise much more slowly than the gross domestic product), and a comparison with the central energy demand variant, IS92a, was made. The energy demand levels used in employing the LESS constructions were in agreement with the opportunities for mitigation that were described in the report.

The dynamics of the global energy system was, however, still described in a very simplistic manner. Even though considerable savings could be achieved through more efficient use of energy than in the present system, there would be costs associated with the introduction of alternative sources of primary energy as well as socio-economic consequences of a markedly different global energy supply system, which would also depend on the *rate of change* that was prescribed. These could not be evaluated adequately. Still, the attempt was interesting and showed that technologies were available. The results, however, were not enough to convice those stakeholders that produce society's energy.

The scenarios of future carbon dioxide emissions as well as the atmospheric carbon dioxide concentrations derived on the basis of these analyses are of course not predictions. They are just scenarios that use different *assumptions* about how the global socio-economic system might change during the next 100 years, which to a considerable degree is largely guesswork. An uncertainty of a somewhat different kind is added to this fundamental one when deriving the climate changes that might result. Global climate models are based on approximations of the fundamental laws of nature and are used in a simplified numerical form to produce projections of future changes of the global climate. However, these basic laws of nature are non-linear in nature and projections are therefore in some respects chaotic. It was not yet possible to account for such complications. Although in a sense interesting, these analyses did not help much at the time in trying to judge whether or not more far-reaching protective measures might already be justified.

8.2.3 *Working Group III assessment report,* Economic and Social Dimensions of Climate Change: *approved in Geneva 25–28 July, and Montreal, 11–14 October 1995*

Working Group III was charged with conducting a technical assessment of the socio-economics of impacts, adaptation and mitigation of climate over both the short and long term and at regional and global levels.

The working group (IPCC, 1996c) had stipulated in its work plan that

> . . . it would place the socio-economic perspectives in the context of sustainable development, and in accordance with the UN Framework Convention on Climate Change to provide a comprehensive treatment of both mitigation and adaptation options while covering all economic sectors and all relevant sources of greenhouse gases and sinks.

This was a tall order for the working group and it was obviously not possible to do more than to begin such efforts, to simply define the scope, structure the task and in this way establish an approach to be pursued. These more general issues were hardly controversial, but when concerned with issues of equity, differences between developed and developing countries and in particular the costs of mitigation, politics easily became part of the discussion and controversies emerged. In addition the cochairmen were not always able to keep a clear distinction between country delegates, on one hand, and representatives from non-governmental organisations, on the other. The difficulties encountered meant that a second session had to be scheduled for October in order to get final approval of the summary for policy makers.

It is worthwhile to first quote a few of the introductory and important sentences of the summary for policy makers from Working Group III. They essentially express common sense and are hardly controversial:

Analyses indicate that a prudent way to deal with climate changes is through a portfolio of actions aiming at mitigation, adaptation and improvement of knowledge. The appropriate portfolio will differ for each country. The challenge is not to find the best policy today for the next 100 years, but to select a prudent strategy and adjust it over time in the light of new information.

Earlier mitigation action may increase flexibility in moving towards stabilization of atmospheric concentrations of greenhouse gases. The choice of abatement paths involves balancing the economic risks of rapid abatement now (that premature capital stock retirement will later be proved unnecessary) against the corresponding risk of delay (that more rapid reduction will then be required, necessitating premature retirement of capital stock).

The literature indicates that 'no-regrets' opportunities are available in most countries and that the risk of aggregate net damage due to climate change, consideration of risk aversion, and the application of the precautionary principle provide rationales for action beyond no regrets.

The values of better information about climate change processes and impacts and society's responses to them are likely to be great. In particular the literature accords high value to information about climate sensitivity to greenhouse gases and aerosols, climate change damage functions, and variables, such as determinants of economic growth and rates of energy efficient improvements.

More heated debates arose in considering the issue of equity and in particular when dealing with equity between generations. This had been discussed extensively in attempts to define sustainable development in meeting '. . . the needs of the present without compromising the ability of future generations to meet their own needs . . .'.[8] The working group also referred to the concept of discount rate frequently used in economic studies for extended periods of time:

Discounting is the principal analytical tool economists use to compare economic effects that occur at different points in time. The choice of discount rate is of crucial technical importance for analyses of climate change policy, because the time horizon is extremely long, and mitigation costs tend to come much earlier than the benefits of avoided damage. The higher the discount rate, the less future benefits and the more current costs matter in the analysis.

In addition, costs for taking action now are much more easily assessed than considering the extent and severity of damage that may be expected many decades into the future. In an exclusively economic analysis a delay of action is therefore often judged as preferable, but other values and social 'costs' also deserve careful consideration. The report concluded that '. . . How best to choose a discount rate is, and will likely remain, an unresolved question in economics.' This also implied that cost–benefit analyses are not as yet an appropriate tool to serve as a basis for political considerations as advocated by for example Cline in his early analyses (Cline, 1992).

Another controversial issue arose when analysing the social costs of anthropogenic climate change. Most studies in the literature concern developed countries but these are sometimes extrapolated and used for analyses in developing countries. The 'statistical value of life' is then often assumed to be different, because of the marked differences in standards of living between developed and developing countries. This may be appropriate when *describing* the way the global economy operates today but is an unacceptable distinction when dealing with the development of the global economy during the next 50–100 years. Striving for sustainable development and an equitable world must be central features of any study of this kind. Nonetheless, a lengthy discussion took place that showed clearly the differences of views between developed and developing countries.

Similarly, the concept of 'purchasing power parity, ppp', which is often used in comparing developed and developing countries, must also be handled with care in order to avoid implicitly made assumptions that are not compatible with the basic goal of a future equitable world.

In attempting to deal with mitigation, costs of response options often become a crucial issue. Analyses have, however, primarily been concerned with developed countries and the working group explicitly avoided making recommendations on policy choices. What matters of course is the net costs (total costs less secondary benefits):

The assessed literature yields a very wide range of estimates of costs and response options. This largely reflects significant differences in assumptions about the efficiency of energy and other markets, and about the ability of government institutions to address perceived market failures and imperfections.

Policymakers should not place too much confidence in the specific numerical results from any one analysis. For example, mitigation cost analyses reveal the cost of mitigation relative to a given baseline, but neither the baseline nor the intervention scenarios should be interpreted as representing likely future conditions. The focus should be on the general insights regarding the underlying determinants of costs.

The summary for policy makers provides a brief overview of the results from some analyses as reported in the literature and aims essentially to provide a general insight. Nevertheless, some quantitative results were also quoted, although the numbers provided were obviously quite uncertain. The following cost estimates summarise available studies at the time:[9]

OECD countries. Top-down analyses suggest that the costs of substantial reductions below 1990 levels could be as high as several percent of GDP. In the specific case of stabilizing emissions at 1990 levels, most studies estimate that annual costs in the range of -0.5% to 2% of GDP could be reached over the next several decades . . . Some bottom-up studies show that the costs for reducing emissions by 20% within two to three decades are negligible to negative . . .

Economies in transition. The potential for cost-effective reductions in energy use is apt to be considerable, but the realizable potential will depend upon what economic and technological development path is chosen, as well as availability of capital to pursue different paths . . .

Developing countries. There may be substantial low-cost fossil fuel carbon dioxide emission reduction opportunities. Development pathways that increase energy efficiency, promote alternative energy technologies, reduce deforestation, and enhance agricultural productivity and biomass energy production can be economically feasible . . . However, these [reductions] are likely to be insufficient to offset rapidly increasing emissions baselines, associated with increased economic growth and overall welfare. Stabilization of carbon dioxide emissions is then likely to be costly.

Although these conclusions from the working group are valuable, *integrated assessments* will be required in order to eliminate at least some of the assumptions that have had to be made in order to deal with the issues with which we are confronted. In the middle of the 1990s attempts in this direction were not yet adequate to increase our understanding of the interplay between the two complex systems significantly, i.e. the global climate system and the global society, of which we all are a part.

8.3 Stabilisation of atmospheric greenhouse gas concentrations

A number of greenhouse gases contribute to the enhanced greenhouse effect, but carbon dioxide is undoubtedly the most important one, even though the role of increasing concentrations of methane should certainly not be ignored. Anyhow a question that quickly comes to mind is: what emission scenarios are permissible in order not to exceed some alternative stabilisation levels?

Intensified studies and increased understanding of the global carbon cycle offered a possible way to address this question (using so-called inverse methods of analysis). About a dozen research groups joined forces and derived sets of scenarios aimed at stabilising atmospheric carbon dioxide concentrations at the levels of 450, 550, 650 and 750 ppmv.[10] Two different pathways towards stabilisation were chosen. In one case mitigation measures were begun at once in order to have a more gradual transition, and in the other, more realistic case, action was delayed and a rather quick transition to a constant concentration level was imposed later (see Figure 8.1). The differences between the two emission paths were quite small and the economically most advantageous one should then obviously be chosen. Table 8.1 shows the total emissions from 1990 to 2100 required to meet these different stabilisation levels.

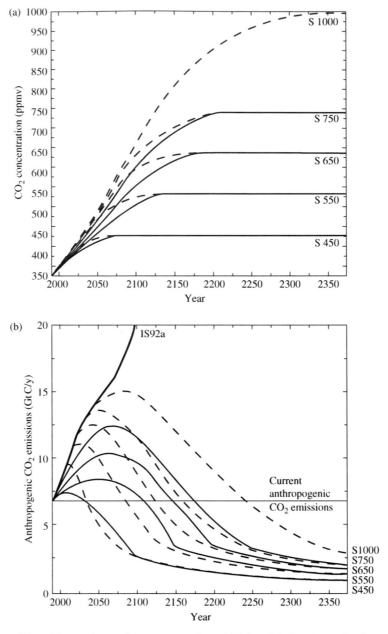

Figure 8.1   Alternative pathways towards stabilising (a) the atmospheric carbon dioxide concentration and (b) the emissions yielding alternative concentration levels in terms of the annual amounts of carbon being emitted. Two courses are shown, one in which there is quick start (solid lines) and another that assumes some initial delays (dashed lines) (IPCC, 1995a).

Table 8.1. *Total emission during 1990–2100 in order to reach stabilisation of atmospheric carbon dioxide concentrations at alternative levels, and implied average annual emissions during the period 1990–2100 (IPCC, 1995a)*

Stabilisation level, ppm	450	550	650	750
Emissions Gt	640–800	880–1060	1000–1240	1220–1420
Annual emissions Gt	5.7–7.3	8.0–9.6	9.1–11.3	11.1–12.9

The ranges express the two path ways towards stabilisation that were run. The uncertainty range is in reality larger, since secondary influences, e.g. due to climate change, have not been considered. Furthermore, stabilisation at 650 and 750 ppm is not yet reached by 2100.

The total accumulated emissions due to fossil fuel burning until 1990 were estimated to have been about 240 Gt C and changing land use had added about an additional 120 Gt C, while the atmospheric concentrations had increased to about 354 ppmv, or in absolute amounts by about 160 Gt C. Total emissions amounted thus to about 360 Gt C, i.e. about 45% had stayed in the atmosphere. The annual emissions from fossil fuel burning in 1990 were about 6.0 Gt C and thus the about 1.1 ton C per capita, but this was very unequally distributed between developed and developing countries. Emissions were increasing quite rapidly, by about 1.5% per year, but there was still no unease about the possibility that the reserves and resources of oil and natural gas might in the long term prove insufficient.

However, neither the magnitude of the net exchange of carbon dioxide between the atmosphere and the terrestrial biosphere nor the uptake by the oceans was quantitatively well known. It had not been possible to account adequately for the fact that only about 45% of the emissions due to fossil fuel burning and the ongoing deforestation in tropical countries actually stayed in the atmosphere. There was a 'missing sink' (see Section 2.2). Any attempt to project likely future concentration levels in the atmosphere as a result of continued human activities therefore required some assumption about the likely future magnitude of the 'missing sink', and the natural starting assumption was of course that it would remain the same. An approximate idea about the constraints that would have to be imposed on future emissions could then be derived. That might, however, be an underestimate of the constraints required.

The annual emissions in 1990 (including the effects of deforestation and changing land use) were about 7.5 Gt C. As can be seen from Figure 8.1 the two lowest stabilisation levels, 450 and 550 ppm would require a substantial reduction of emissions, in the former case beginning within just a few decades and for 550 ppm towards the middle of the twenty-first century.

The inertia of the carbon cycle and therefore the need for substantial decreases of the emissions to achieve stabilisation of concentrations seemingly impressed country delegates to the IPCC and similarly so the country representatives to the INC (and later also the subsidiary bodies of the Convention). It would become a central issue during the next two years when preparing for the conference of the parties to the Climate Convention scheduled for Kyoto.

The incomplete and rather uncertain projections of the regional characteristics of the expected change of temperature and precipitation during the next several decades were still a stumbling block for progress in persuading countries to act rather than continue negotiations. The fact that some climate change indeed seemed inevitable had not yet convinced countries that action had to start soon in order to keep the forthcoming climate change within bounds. Other political issues on the global scene were obviously judged to be more important.

8.4 The synthesis report

Following a resolution adopted by the executive council of the WMO in July 1992 and in consultation with the cochairmen of the three working groups, I proposed to the IPCC Bureau that a synthesis report should be produced. It was important that a well-prepared outline for such an effort was agreed early. Care would also be required in the analysis to avoid unnecessary controversies. Political emphasis had repeatedly been placed upon an analysis of the objectives of the Convention as given in Article 2. This had also been the key issue for the workshop in Fortaleza in October 1994. It therefore seemed logical to focus the synthesis report on providing scientific information relevant for the political analyses of this issue by country delegates to the Climate Convention. I also recalled the disappointment shown by the INC chairman when insufficient information was available for a more effective pursuit of this issue by the INC. The IPCC contribution was based on the assessments by the working groups but with no recommendations for action, which was left for the country delegates to the Convention.

A special working party was formed, including the six cochairmen of the three working groups as well as a number of other experts.[11] It should be noted that several of the lead authors engaged in the work on SAR were included, as well as external members. In retrospect, it may perhaps be said that the committee was unduly dominated by Anglo-American researchers. It was further decided that the regular IPCC review procedures should be used before submitting the draft of the report to the IPCC plenary session for consideration. A first meeting of the working party took place in Athens at the beginning of May 1995 and lasted several days. The group met altogether on three occasions in the course of the year.

The first draft of the report was circulated for review in August–September. In the process of writing, the short and succinct report that I had envisaged had expanded into a more extensive document and obviously a summary for policy makers was needed in order to have a document that could be approved line by line at the IPCC plenary session in December. This caused feverish activity amongst some key lobby organisations and stakeholders in the USA that resulted in a long letter to the US State Department, US Department of Energy and US Environmental Protection Agency.[12] The letter ended with the request that '. . . you will agree with us that the approach taken to develop and consider the proposed Synthesis Report is fatally flawed. We look to our government to use its leadership position in the IPCC to obtain corrective action.'

It was further explicitly requested that there should be no summary for policy makers, to the synthesis report, but rather that the full report should be approved line by line at an extraordinary IPCC plenary session in April 1996 and that the deadline for submitting comments to the synthesis report should be extended by a month until October 16. '. . . Unless these procedural issues can be resolved properly, there should be no Synthesis Report, which is neither necessary nor prudent in any event.'

No action was taken by the US Administration. In fact, I was asked to discuss climate change matters with a small group of people from the ministries concerned, and in the evening I was invited to dinner by Assistant Secretary of State, Tim Wirth in Washington, DC.

The final synthesis report was short (15 pages) and succinct.[13] It was discussed extensively during the last days of the Rome session and the procedure was agreed that if a unanimous view on an issue could not be reached, a footnote should be inserted in the final report referring to the country or the minority of countries that opposed the conclusion. In fact this procedure was used on only a few occasions. The final discussions of the report took place between eight o'clock in the evening and midnight on the last day of the session, at which time we had to end our discussions for logistic reasons. The modifications of the draft were actually minor. Perhaps the kind arrangements of our Italian hosts who served an impromptu supper, consisting of spaghetti and meat sauce and red wine at the conference table, contributed to the successful conclusion of the session. However, the main opponents of the first draft had left the session before I finally concluded it at midnight. In fact, this synthesis report has been repeatedly referred to during the following years and has contributed in an important manner to the advancement of the negotiations.

The key task of the synthesis report was to set the stage for what was needed in order to address Article 2 of the Convention:

> ... the ultimate objective of the Climate Convention is ... stabilization of greenhouse gas concentrations in the atmosphere at a level that would prevent dangerous anthropogenic interference with the climate system, ...

and the Panel added

> ... and the charting of the future, which allows for economic development that is sustainable. The purpose of the synthesis report is to provide scientific, technical and socio-economic information that can be used in addressing these challenges. It is based on the 1994 and 1995 reports of the IPCC Working Groups.

9

In the aftermath of the IPCC second assessment

The IPCC second assessment is severely criticised but it still plays an important role in preparing for the third conference of the parties to the Climate Convention scheduled for December, 1997, in Kyoto.

9.1 The post-Second Assessment Report discussions of an action programme to be agreed in Kyoto

The year 1995 ended with the completion of the SAR at the eleventh IPCC session in Rome. Earlier in the year the Climate Convention had formulated a mandate in Berlin for negotiations to be conducted during 1996 and 1997 in order to arrive at decisions on actions to be taken at the third conference of the parties at Kyoto. But the journey that awaited the IPCC during the next two years was not to be a smooth one. Even though the scientists that had been involved with the IPCC supported the conclusions that had been drawn, there was not complete unanimity in the scientific community about all that the IPCC had concluded, which of course was no surprise. Scientific progress is always based on challenging current knowledge. In addition stakeholders organised themselves in order to protect their conceived economic interests. The IPCC bulk reports exposed the uncertainties and disagreements as well as possible wherever they appeared, and the summaries for policy makers expressed these conclusions carefully in order not to give a false impression of certainty, where there was uncertainty. The paragraph that summarised our understanding of the degree of human interference with the climate system may be taken as an example. It was concluded, *fully recognising the uncertainty in a caveat* (see Section 8.2.1), that '. . . the balance of evidence suggests that there is a discernible human influence on global climate.' As I viewed the available knowledge base at the time, this expressed carefully and well the opinion that the conclusions were not yet final and firm, but that human interference was likely to contribute significantly to the ongoing change, but by how much was not possible to assess. However, the preceding sentences that expressed the reasons for uncertainty were seldom quoted, which led to misunderstandings and accusations that the summary for policy makers was not an appropriate summary.

It had early been agreed that the IPCC should focus on the compilation of available scientific knowledge, but not express views on whether or not one or other conclusion regarding a changing climate implied a threat to the world and the well-being of people on earth, i.e. whether or not it was 'dangerous', as formulated in the Climate Convention.

Words such as catastrophe, apocalyptic change, threat to mankind, etc. are not found in the IPCC reports. Accusations from critics and lobbyists that the IPCC was not scientifically objective in this regard were indeed inappropriate and should be seen as their way of attempting to discredit the scientific process to their own advantage, thereby avoiding a serious discussion with the scientific community. Most people did not have access to the full reports and seldom checked what was quoted from the IPCC reports and how those quotations were chosen. It is also noteworthy that the critics had seldom, in many cases never, published scientific papers in the peer-reviewed scientific literature that covered the relevant fields of research and therefore often misinterpreted what had been said, occasionally even in the summaries for policy makers.

In any case the IPCC conclusions were criticised. Sometimes there would be cautiously expressed doubts that the IPCC conclusions were not fully justified in the light of the major uncertainties and the numerous assumptions that were made in the development of climate change simulations. On other occasions the critics were much harsher, accusing the scientists and the IPCC leadership, in particular, of manipulations and using dishonest procedures.

However, the completion of the SAR gave an impetus to the work of the Climate Convention, and the ongoing negotiations in the ad-hoc group on the Berlin mandate in particular. I continued to address the key bodies of the Convention and now had much to tell them about, i.e. the conclusions reached in the SAR.

9.2 The IPCC assessment is challenged

9.2.1 Some basic scientific issues arise

The IPCC assertion that human-induced global warming was on its way was challenged immediately when the conclusion of the IPCC second assessment became known, with the simple argument that the climate had always been changing. Were the changes observed during the twentieth century really not just a continuation of the recovery from the little ice age during preceding centuries? This view was revived when Christy and his colleagues showed with the aid of satellite observations that the temperature in the upper troposphere, well above the earth's surface, had decreased in recent decades contrary to what observations at the earth's surface indicated.[1] Interpretation of the data was difficult, however, because the possible long-term changes were obscured by temporary variations, for example as a result of volcanic eruptions (El-Chinon in 1983 and Pinatubo in 1991). On the other hand, it was generally agreed that the temperature in the lower stratosphere was decreasing slowly as theoretically expected because of the increasing concentration of carbon dioxide. It was, however, soon pointed out by Hurrell and Trenberth[2] that errors had been introduced when merging time series obtained from different satellites. They concluded that in the upper troposphere '. . . the trend in the satellite measurements is therefore likely to be positive, albeit small.' In addition, Wentz and Schabel showed that a slow change of the satellite's orbit might also influence the temperature measurements. It should further be recognised that in order to draw trustworthy conclusions the measurements required a precision of better than $0.1\,°C$, which is not easy to achieve in satellite measurements extending over a decade or two. It has later also been recognised that there were systematic errors in the earlier direct temperature observations from balloons in the upper troposphere and lower stratosphere because of inadequate shielding of the sensors from solar radiation. This showed up as a systematic difference between day and night measurements. Therefore the observed decrease might have been due to erroneously high temperatures measured from balloons before satellite observations became available, i.e. during the 1960s and early 1970s. In 2007 there are hardly any internal inconsistencies in the changes of the vertical temperature profile as deduced by different means. This has been resolved by careful scientific analyses. However, this issue was discussed extensively during the preparations for the conference of the parties of the Convention, scheduled to take place in Kyoto at the end of 1997 (see Christy *et al.* (2000)).

9.2.2 Objections from industry and its supporters

Efforts to discredit the assessment process were intensified in 1996. US industrial groups were regularly represented as non-government organisations when the IPCC met, as well as at meetings of Convention bodies when this was allowed. Non-government organisations were sometimes given the opportunity to present their views, but on the whole IPCC sessions were the place for constructive discussions between government representatives, scientific experts and the IPCC leadership. However, some representatives from oil-producing countries as well as others were obviously occasionally given directives from home to protect national interests by challenging the scientific conclusions reached by the IPCC working groups. This seldom influenced the reports, but occasionally led to more clarity in the way in which the conclusions were expressed. In addition, industrialists were very active in their respective countries in communicating their views regarding global climate change to the public, politicians and non-government organisations representing environmental groups.

A more organised opposition to the IPCC's conclusions began in the USA on Earth Day (22 April 1996), with a message distributed widely, including to every member of the US Congress, and with the first issue of the *State of the Climate Report* attached in which the IPCC conclusions were challenged.[3] However, just as this report was about to be published, the Union of Concerned Scientists denounced it in a press release, based on earlier contributions to the media debate about global warming by the man in charge, Patrick Michaels: 'The forthcoming climate change report sponsored by the Western Fuels Association is like a lung cancer study funded by the tobacco industry.' Rather than using such language my responses on a few of Michaels' contributions in this first issue of the report were as follows. He had written

In 1992, during the apocalyptic heydays, President George Bush ratified the Framework Convention on Climate Change, a UN treaty that permits any 'consensus' of nations to dictate the internal energy policy of other sovereign states.

The scientists had certainly not contributed to the 'apocalyptic heydays' and his way of describing matters was hardly a scientific analysis. Further, in its first issue *The State of the Climate Report* systematically misquoted and ridiculed the efforts by scientists in 1991–2, particularly those of John Mitchell and his colleagues at the Hadley Centre in the UK, to provide the best possible evidence for the Climate Convention negotiations. The early climate models did not include the role of aerosols, since this was first generally recognised to be of serious concern in 1991, which of course meant considerable uncertainty in any modelling effort. But the uncertainty of the estimated sensitivity of the climate system was indeed considered in the IPCC report from 1992 to the extent it was known at that time (see IPCC (1992b)). In addition the articles that Michaels reviewed or referred to were very selectively chosen.

The way Michaels dealt with the climate change issue in the first issue of the *State of Climate Report* disqualified him from taking part as a serious fellow scientist in the climate change debate. His statements were simply not trustworthy. Nevertheless, his writing was widely distributed and influenced public opinion.

On 12 June 1996, Professor Frederick Seitz published in the *Wall Street Journal* a vicious attack on the way the IPCC procedures had been applied at the IPCC Working Group I plenary session in Madrid in November 1995, at which the IPCC SAR was considered for final approval:[4]

In my more than 60 years as a member of the American scientific community, including services as president of the National Academy of Sciences and the American Physical Society, I have never witnessed a more disturbing corruption of the peer-review process than the events that led to this IPCC report.

Benjamin Santer, one of the convening lead authors of the chapter of the SAR in focus, responded to this assault and was supported by 40 fellow scientists, some of whom were coauthors of the chapter that was the subject of the controversy or of other chapters in the IPCC SAR. However, their names were deleted without the readers of the *Wall Street Journal* being informed. Some quotes from his letter (which was much shortened) reveal the strong reaction from a number of scientists that were engaged in the IPCC assessment.[5]

Frederick Seitz' June 12 editorial page piece, 'A major Deception on "Global Warming" ' wrongly accuses both the IPCC and a member of the climate science community of violation of procedure and deception. Not only does he thereby demonstrate ignorance of both topic and the IPCC process, but his actions reflect an apparent attempt to divert attention away from the scientific evidence of human effect on global climate by attacking the scientists concerned with investigating this issue.

There has been no dishonesty, no corruption of the peer-review process and no bias – political, environmental or otherwise. Dr Seitz claims that the scientific content of Chapter 8 was altered by the changes made after the Madrid meeting. This is incorrect. The present version of Chapter 8 draws precisely the same 'bottom line' conclusion as the original October 9th version of the chapter. Taken together, these results point towards a human influence on climate.

Dr Seitz is not a climate scientist. He was not involved in the process of putting together the 1995 IPCC report on the science of climate change. He did not attend the Madrid IPCC meeting on which he reports. He was not privy to the hundreds of comments received by Chapter 8 Lead Authors before writing his editorial. More seriously, before writing his editorial, he did not contact any of the Lead Authors of Chapter 8 in order to obtain information as to how or why changes were made to Chapter 8 after Madrid.

The University Corporation for Atmospheric Research, UCAR, and the American Meteorological Society strongly supported the intervention by Santer and his colleagues, and reproduced the exchanges of letters in full in the *Bulletin of American Meteorological Society*,[6] i.e. including the parts that had been deleted by the *Wall Street Journal*.

I also intervened jointly with the cochairmen of Working Group I with a letter to the *Wall Street Journal* in support of Ben Santer, primarily in order to describe in some detail the procedures adopted by the IPCC, but the letter was similarly severely edited. Less than half of it was published in its original form. The crucial part that explained the IPCC review process, which was the basis for Seitz's attack, was omitted. It is in this context interesting to note that the US State Department, in responding to the IPCC's request for comments to the 9 October 1995 draft of the summary for policy makers, had attached a letter to the cochairmen of the IPCC Working Group I (dated 15 November 1995) that emphasised

In keeping with past practice in Working Group I, it is essential that the chapters not be finalised prior to completion of discussions at the IPCC working group I plenary, and that chapter authors modify the text in an appropriate manner following discussions in Madrid.

This paragraph describes precisely the procedure that had been adopted and used and that was the key issue in Dr Seitz's accusations. As described in the previous chapter the bulk report on which the summary for policy makers is based, was accepted unanimously by the working group in Madrid. The two convening lead authors agreed that the modifications of the summary for policy makers approved by the working group were consistent with the findings of the lead author team after consideration of all the written comments that had been submitted and the discussions at the working group meeting, and that modifications of the bulk report were made in order to ensure internal consistency. The cochairmen of the working group were responsible for ensuring that the appropriate changes were made.

This controversial issue also resulted in two letters (dated 30 May and 26 June), being sent to me, one from the Global Climate Coalition (John Schlaes) and the other from The Climate Council (Donald Pearlman). Copies of these were also sent to ten key members of the US Congress as well as the Advisor for Science and Technology and Assistant to the US President (John Gibson), and the Assistant Secretary of State (Eileen Clausen). The second letter spelled out in detail the ignorance of the rules of procedure in Madrid as viewed by these two stakeholder groups. They claimed that the promise in an earlier statement of mine that '. . . the lead authors had overall responsibility for the background documents for ensuring that review comments were properly considered and that the reports presented a comprehensive, objective and balanced view of the areas they cover . . .' had not been fulfilled. The IPCC was accused in very strong terms of breaking its own rules of procedure in Madrid and it was claimed that Chapter 8 of Working Group I report seriously lacked the balance that is so essential for an objective assessment of the climate change issue. The letter ended with a quotation from an editorial in *Nature* (June 13, 1996) that was implicitly supporting their accusations:

On a topic where political sensitivities run high, the integrity of the reviewing and approval process is, as in any scientific publishing endeavour, an essential element in assuring the credibility of the resulting conclusions.

Of course in principle I share the sentiments expressed in the comments by *Nature*, but the IPCC's response to Dr Seitz, as described before, is also relevant here. The IPCC's view was expressed in a detailed response to the two stakeholder groups by me and the two cochairmen of Working Group I, and copies of the response were sent to those that had received a copy of their letter. It is interesting to note the influence of the lobbying groups even on a journal such as *Nature* which it is hoped should present an independent and carefully balanced view of matters of this kind. I was never approached by the journal to give the IPCC's views.

The comments in the letter from Dr Seitz and his colleagues and the views expressed by Mr Shlaes and Mr Pearlman are formalistic and obviously aimed at impeding the cooperation within the scientific community that was necessary for the IPCC assessments. A very stringent adherence to the rules of procedure would severely inhibit the regular and open exchange of views that is so vital for the scientific discourse. Further, the responsibility for the final versions of the bulk reports according to the rules of procedure rests with the lead authors. They are supposed to exercise their scientific judgment and the final version of the report is merely accepted, not approved, at the working group plenary session. The sizes of the lead author teams and the extensive review process ensure that the best possible unbiased outcome is achieved. I judged that the review comments had been incorporated properly, both those that had been submitted in writing and those that were put forward during the plenary session in Madrid, and that the final version of Chapter 8 expressed the views of the lead author team as a whole in a balanced way. In fact, the final version of the chapter is quite close to the version that was submitted to delegates on 9 October. The arguments for a discernible climate change had become somewhat more convincing. There was no hidden agenda behind the modifications that were made. Uncertainties were emphasised and can be found in the chapter itself, and support was given to the message as formulated in the summary for policy makers. The concluding sentence in the crucial paragraph of the bulk report was rewritten because of discussions at the working group meeting that expressed the view that '. . . *an as yet not quantifiable, but still discernible, part of the ongoing climate change could be attributed to human activities* . . .' The crucial sentence in the summary for policy makers is almost identical. Nevertheless, the modifications made in Chapter 8 were seized upon in spite of the legitimacy of the procedure that had been followed. As we shall see, the essence of these IPCC conclusions was very well upheld in the third assessment about five years later.

The controversies were, however, not yet over. Frederick Seitz, in his capacity as president of the George C. Marshall Institute in Washington, DC, assembled the small group of sceptics from among the institute's leaders and acquired support from some senators in US Congress.[7] They wrote a letter to the two cochairmen of Working Group I and myself (dated 20 August 1996) and to Tim Wirth at the US State Department, again challenging the outcome of the Madrid meeting. On this occasion the politics of climate change was more in focus. Some of the senators who had signed the letter had attended the second conference of the parties to the Climate Convention in Geneva in July as observers.

The response from the State Department (dated 24 September 1996) was quite detailed and succinct. A short and carefully written review of the relevant scientific conclusions in the IPCC SAR was given (presumably prepared by Bob Watson, the cochairman of Working Group II and in the USA responsible in the White House for environmental issues). Wirth rejected the accusations and then sketched the Administration's view of the US policy that should be aimed for during the next few years. This was forward looking and stressed the need to consider the economic issues that might arise. On such matters he saw close cooperation with industry as essential but also emphasised the necessity of considering the climate issue as part of striving for long-term global sustainability.

IPCC also responded to Seitz's letter. It is appropriate to quote some key statements in this response:

The IPCC Working Group I meeting in Madrid . . . was a scientific meeting which discussed scientific information and scientific presentations. Political issues were not raised or discussed at the meeting and its conclusions were not compromised by any political considerations . . .

We wish to emphasise that the Summary for Policymakers is a balanced scientific document that was unanimously agreed in Madrid by delegates from 96 countries, amongst which were countries with very widely differing political agendas – a good number of the OPEC countries, for instance were represented. Since the Madrid meeting, none of these countries has challenged either the procedures followed or the changes made to the scientific chapters in response to the decisions at the Madrid meeting. . .

A further point you raise concerns the procedures for taking account of the very large numbers of review comments on the scientific chapters. These are considered, not just by the convening lead author of each chapter but by all members of the team of lead authors, typically between 3 and 10 (there were 4 for chapter 8), who come from different scientific institutions and, so far as possible, from different parts of the scientific area considered in the chapter. It is the lead authors acting as a team who are responsible for the overall response to the review comments.

One additional comment might be of interest. Professor Seitz proposed in a letter to me (personal communication) shortly thereafter that I should write to Tim Wirth and emphasise that IPCC in its last assessment has expressed the view that

. . . to date, pattern-based studies have not been able to quantify the magnitude of the greenhouse gas or aerosol effect on climate. It is therefore inappropriate and incorrect for the Ministerial Declaration on July 18, 1996 to link the conclusion about 'human influence' to a temperature increase of 2 °C by 2100 . . .

He even drafted a letter for me, which I was supposed merely to sign. I have no comments on this strange approach other than saying that the issue is much more complex than can be expressed by the wording of one single sentence and I was indeed amazed about Dr Seitz's way of proceeding.

9.2.3 The US Administration and Democrats in US Congress support the IPCC assessment

In the midst of these controversies the second conference of the parties to the Convention took palace in Geneva. There is not much to report from this session except the exchanges of views I had with the two key representatives from the USA, Tim Wirth and Eileen Clausen. The following quotation from Tim Wirth's statement to the Conference might, however, also be of interest:[8]

In our opinion, the IPCC has clearly demonstrated that action must be taken to address this challenge and, as agreed in Berlin, more needs to be done through the Convention. This problem cannot be wished away. The science cannot be ignored and is increasingly compelling . . . Unhappily [however], while the established international scientific process is working well, the international policy process, as established under the Convention, has not been as successful.

The United States is committed to making the international climate change process work. The science is convincing; concern about global warming is real and we must continue to take steps to address this problem consistent with our long-term economic and environmental aspirations. Working together, it is imperative that we marshal the creativity and will necessary to address this far-reaching challenge. The United States hopes we can negotiate an agreement that is comprehensive, flexible, fair and certain, and which will help prepare our country and the world – environmentally and economically – for the next century.

It is very clear from this statement that the US policy on climate change would be built on the IPCC assessments, but with due consideration of US economic interests and the importance of international trade. There was obvious tension between the Administration and a sizable part of US industry.

In October 1996 George Brown, a Democratic member of the US House of Representatives and member of the Committee on Science, distributed to the Democratic caucus of the Committee, a detailed analysis of the role of science in politics.[9] A quotation from his circular to the Committee is of interest:

Three major hearings, entitled Scientific Integrity and Public Trust, were convened by the Energy and Environment Subcommittee to showcase allegations that science had been distorted to promote an environmentalist agenda, The hearings focused on alleged abuses in the science on stratospheric ozone depletion, global climate change, and the health risk posed by dioxin. The series of hearings did not occur in a vacuum. The new Republican Congress promised as part of its 'Contract with America' to fundamentally change the way environmental regulations would be promulgated. With a new majority [in Congress] the Republicans launched an attack on the basic methods by which environmental regulations could be established.

Several of the issues that had been criticised by sceptics were brought up in the hearings that were held in order to establish the factual background for a closer analysis. Dr Michaels, as editor of the *State of the Climate Report* was in the forefront defending the sceptics' views. A report was prepared by the staff of the Energy and Environmental Subcommittee, based on analyses of a number of case studies that had been presented at the hearings, to serve as a basis for the judgements by the committee members. Some of its conclusions deserve attention:

[Climate] models have been improved markedly . . . and describe the characteristics of the climate reliably enough to support policy decisions. The arguments 'du jour' advanced by the sceptics, who seek to exploit these remaining uncertainties, are generally invalid, because the purported comparison set up a false test between incomparable data and specific model simulations . . .
    . . . the alleged discrepancy between GCMs [global climate models] 'predictions' and satellite data showing no warming over the last 17 years is another inappropriate comparison which also fails to prove that the GCMs are wrong. Dr Michaels' claim that the limitations of the GCMs were deliberately understated is not substantiated by the IPCC documents' . . .
    . . . Dr Michaels' request for data [from the UK Meteorological Office] may be appropriate, and access to such data could conceivably contribute to the review process. However, there is no evidence that the denial of such data [that Michaels had experienced] was inconsistent with US law, or was part of an overall conspiracy to suppress dissenting views.

Dr Mahlman[10] had testified to the question: 'Is there a scientific basis for the claim that humans are responsible for global warming?', and answered 'Global surface warming over the past century is virtually certain. The observed warming in the surface temperature records of about one degree Fahrenheit cannot yet be unambiguously ascribed to greenhouse warming. However, no other hypothesis is nearly as credible.' It is indeed satisfying that quite an extensive and careful analysis of the issue of scientific integrity and public trust is given this attention.

Another set of incidents deserves some attention: Fred Singer (a member of the Marshall Institute) toured Europe in June 1997 giving lectures in London, Munich, Copenhagen, Helsinki and finally in Stockholm. Naturally, I was the target during his visit to Sweden. He spoke to a rather limited audience, and his presentation was followed by a brief and unbalanced discussion because only a very select audience had been invited to attend his presentation and most of those present had limited knowledge of the science of climate change.

On the following day Singer issued a press release that was effectively distributed through Greenwire, in which he reported on the meeting and about his own presentation as if the report had been given by an unbiased observer. It is a remarkable document, describing the outcome of the meeting in a very subjective manner. It was solely aimed at discrediting the second IPCC assessment with arguments that were seldom based on the current literature and most of which had been refuted by the IPCC in its second assessment. His account of what happened at the meeting is inaccurate and misleading. The discussion after his lecture was the starting point for contradicting the IPCC's view that the issue of global warming had to be taken seriously. Singer's tainted arguments that then followed show his approach in arguing against the IPCC analyses. The points were neither significant nor verifiable.

I replied that we cannot directly associate the occurrence of specific extreme events with the ongoing global warming. However, taking some first steps to prevent a possible future global climate change should be seriously considered in the light of available knowledge about the *global change issue as a whole*, as had also been emphasised in the IPCC SAR.

Singer's press release made headlines particularly in the USA, but also elsewhere. I was amazed that this serious issue was presented with an almost complete lack of credible scientific analysis, but Singer undoubtedly gained considerable publicity which spread a sense of uncertainty that had some, though still limited, political impact.

9.2.4 The IPCC second assessment causes fewer controversies in Europe

The EU was responsible for responding to the IPCC assessment and coordinating the joint actions of the 15 member states in the broader international context. The need to take action soon was largely agreed upon politically without much controversy. The prospects of reducing emissions were reasonably good in a first round. In the UK coal was being replaced by gas from its assets in the North Sea. France continued to expand its use of nuclear energy for electricity production and Germany was replacing inefficient power production plant in the former East Germany with modern equipment. However, industry was not ready to accept quick action because of the capital cost involved in the early retirement of existing production units and the loss of competitiveness. Politicians also realised that the rapidly expanding transport sector meant an increasing demand for fuel. The reluctance to take action was sometimes expressed by casting doubts on the IPCC analyses.

In the UK the BBC arranged a mock trial in which the representatives of the IPCC were accused of causing future damage to industry because of the protective measures that would have to be taken, while on the other hand those not willing to accept such measures were accused of contributing to future damage due to a human-induced climate change. The criticism of the IPCC conclusions was led by Peter Emsley, a professor of physical chemistry. Professor Richard Linzen had been called in from the USA, while I and a few other IPCC people were defending the IPCC analyses. There were prosecutors and lawyers as in legal procedures. I did not really appreciate the idea behind this setup. It seemed strange to judge whether scientific analyses should be accepted or rejected in terms of scientists being guilty or not guilty of future damage, particularly when there were as yet few, if any, reliable risk analyses.

Professor Aksel Wiin-Nielsen in Copenhagen, former head of the European Medium Range Weather Forecasting Centre in Bracknell, England, and also former secretary general of the WMO (Geneva), had major objections to the IPCC conclusions and even proposed that the IPCC should be dissolved. I never understood why he became an opponent of the IPCC efforts, when he had made well-recognised scientific contributions in the past. His very simplistic scientific analyses regarding the issue of human-induced climate change were not supported by adequate data. His criticism was not taken seriously by former colleagues and other scientists who had taken part in the IPCC assessment.

On the other hand, Klaus Hasselmann, director of the Max-Planck Institute for Meteorology in Hamburg and himself seriously engaged in the climate change issue, gave strong public support to the IPCC view that a human-induced global change of climate was on its way and that this conclusion was plausible with a 95% probability. The final paragraph of his article in the major German paper *Die Zeit* reads (my translation):[11]

The difficulty of handling the climate issue, politically as well as publicly, is the necessity of accepting a long-term perspective. If this were possible, one would also realise that a solution of the climate change issue can be found. A long-term, steady policy of small steps that would not be harmful to industry, and simultaneously a gradual transition to the use of renewable sources of energy, would be required. Kyoto would be a success, even if to begin with only small steps were taken (nor can more than this be expected), provided agreements could be reached that more stringent measures during later years would be implemented gradually.

This was of course in sharp contrast to the views held by Peter Dietze in Munich. He developed a model for the carbon cycle that he used extensively in criticising the contemporary literature in the field. He ignored the fact that the IPCC's task was to bring together and analyse in a critical manner the studies that had been published in scientific journals. He never published any analyses himself, because either he simply did not submit any for publication or those that he tried to have published were rejected in the peer-review process. He established a network of sceptics on the web, but he soon lost the trust of scientists in general because of his lack of basic knowledge of the carbon cycle and the climate system, and also because he ridiculed those scientists who tried to tell him their views of his conclusions.

It is amazing to see the marked differences between, on one hand, the careful efforts of the scientists serving the IPCC in the assessment of our knowledge about the gradual change of climate, based on the scientific literature that was available, and on the other, the almost always scientifically inadequate approaches in the shallow analyses by sceptics who lacked the scientific knowledge to deal with the climate change issue. In addition, the sceptics disregarded the socio-economic issues, such as those due to increasing exploitation of natural resources and their implications for future development, as well as the major differences between developed and developing countries and between the rich and the poor.

9.3 Preparations for the third conference of the parties to FCCC in Kyoto

9.3.1 Focusing on reducing emissions and stabilising greenhouse gas concentrations

The two Convention committees SBSTA and AGBM met about three times a year in order to develop a protocol to be attached to the Convention that would prescribe country commitments. I attended a number of these meetings to present those parts of the SAR that were particularly relevant to the ongoing discussions. My report to the second session of SBSTA in February 1996, was rather detailed and focused on the issue of stabilising atmospheric greenhouse gas concentrations (with particular emphasis on the inertia of the climate system), and the importance of accounting for the influence of aerosols. I stressed my view that the changes that might be needed in global society, particularly with regard to future energy supply systems, would not come about easily and quickly (IPCC, 1996a).

Many of us in the IPCC Bureau felt quite strongly that the process for the completion of a full assessment report was long and cumbersome, even though we also realised that the process was essential in order to preserve the integrity of the IPCC. The idea emerged of producing simple short papers about key issues, now that serious negotiations about mitigation had begun. The IPCC accordingly agreed that a set of technical papers should be produced and that four would be available before the third conference of the parties to the Convention.[12] Three of these were directly related to the ongoing negotiations of a protocol:

1. Technologies, policies and measures for mitigating climate change
2. Stabilisation of atmospheric greenhouse gases
4. Implications of proposed CO_2 limitations.

The choice of subject matter was made after consultations in the joint working group of the convention and the IPCC, and it was also agreed that the content should be based on the SAR, so that the reviews and final acceptance of the reports would be simplified.

Agreeing the approach to be followed in Technical Paper 4, however, caused some controversy. A number of countries had suggested to SBSTA desirable reductions of the emissions, and the IPCC intention was to bring together the implications of these alternatives in a useful manner.[13] The IPCC's view was naturally that the more alternatives the better, since this would most clearly show a range of possible future emissions. However, this became politically controversial because the analysis of any particular emissions scenario could be taken as an implicit recommendation that this choice would be preferable. For example, Saudi Arabia and Nigeria argued forcefully for scenarios that would not hamper the activities of oil-producing countries. It therefore became very important that the IPCC should decide independently what to do. It was agreed that the paper would compare several alternatives. However, there were few studies available that also provided information about the economic and other implications of the different emissions profiles that were chosen.

The SAR clarified the limitations of future global emissions that were required in order to stabilise the atmospheric carbon dioxide concentration, the lower the stable concentration aimed for, the more severe the limitations necessary. The following short and somewhat simplified account summarises the gist of the message.[14]

In order not to exceed carbon dioxide concentrations of 650, 550, or 450 ppmv the global emissions due to fossil fuel burning must not rise from 6.0 Gt C in 1990 to above about 11, 10, and 8.5 Gt C per year respectively, by about the middle of the twenty-first century, and thereafter they need to decrease to well below the 1995 emissions in the latter part of the century. The lower a choice of the stabilisation level assumed, the earlier this decline would have to begin.

The world population was at the time about 5.5 billion (10^9) (and expected to increase to above 10 billion towards the end of the twenty-first century). Thus the annual global per capita emissions in 1995 were about 1.1 ton C. However, the differences between countries were large: the emissions were on average about 3.1 ton C per capita in developed countries, but merely about 0.55 ton C in developing countries, i.e. a factor of about 5½ less. Annual per capita emissions from individual countries differed even more. USA emissions were about 5.2 ton C per capita, while India's were about 0.3 ton C per capita.

Simple estimates then tell us that the average annual global per capita emissions must never increase beyond 1.1–1.3 ton C, i.e. not much above what had already been reached in 1995, if more than a doubling of the atmospheric concentration was to be avoided. It also follows that even if developed countries were to decrease their per capita emissions to one third of those in 1995 (i.e. to about the prevailing *global average in 1995*), developing countries would on average never be able to emit much more than about twice their 1995 per capita emissions, if concentration levels in excess of 550 ppmv were to be avoided. This would mean that towards the end of the twenty-first century developed and developing countries would be emitting about the same amount per capita. In other words, during their process of industrialisation developing countries would never be able to use fossil fuels to provide energy in the manner developed countries did during the latter part of the twentieth century. This was obviously going to be a crucial dilemma in the next few decades. At the time, this quite severe constraint on the future use of fossil fuels for energy production was not really appreciated, and in some cases not a recognised at all, particularly so in developing countries.

Technical Paper 4 was not intended to be regarded as recommendations for the future, but was meant to illustrate the difficulties that would be encountered if stabilisation at 450 or 550 ppmv was attempted. Technological development aiming at increasing energy end use efficiency and employing renewable primary energy sources was therefore essential. The unease amongst developing countries became very obvious, however, when the draft of Technical Paper 4 was circulated for review. Gylvan Meira, cochairman of Working Group I, and a lead author of the paper, and I received a strong letter of complaint from China.[15] The gist of the letter becomes clear from the following excerpt:

. . . According to the analysis by this Paper, it seems to be 'rational' enough for the developed countries to reduce only 1% or 2% of the emissions [per year], though they already have a large portion of the emissions . . . in order to achieve stabilisation at 450 ppmv, the annual total emissions of all developing countries could not possibly exceed 2.0 Gt C . . . This is totally irrational. Even if the emissions stabilise at 650 ppm, the annual emissions of all developing countries will be less than 6.5 Gt C . . . We feel sorry for such a scientific assessment lacking fairness and equity, and we firmly request to revise this Technical Paper.

The paper had obviously been misunderstood and was modified, but the key message remained and so also the misunderstanding. The Chinese delegation raised the issue again at the seventh session of the SBSTA in October, 1997, referring strongly to the lack of fairness. I had the opportunity to discuss the matter personally with the principal delegate from China, Mr Zou, at an evening reception at the Chinese Embassy, but was not able to resolve the misunderstanding, which may indicate that the letter had been part of a deliberate effort to strengthen China's leading role amongst developing countries. I wrote a long letter to clarify the inability of the natural global environment to respond quickly to the disturbances that human activities were already imposing, particularly the burning of fossil fuels.[16]

This was actually my final appearance at an SBSTA session and my attention was still on the outcome of the thirteenth session of the IPCC (see IPCC (1997e)). I was, however, most anxious to refocus attention of country representatives at the session on the key issues that I had emphasised repeatedly in the past, not least the stabilisation of atmospheric carbon dioxide concentrations already referred to above (see IPCC (1997f)).

The inertia of the climate system and the emissions of aerosols to the atmosphere imply that probably less than half of the greenhouse effect due to past emissions of greenhouse gases as yet has been realized in terms of global warming. Since these gases are long-lived, the full effects of past emissions will ultimately show up even if and when global emissions are reduced to stabilize greenhouse gas concentrations in the atmosphere. This general and relatively slow response of the climate system to external perturbations . . . also means that measures taken in order to stabilize or decrease emissions will show up only slowly. It is most essential to recognize this characteristic feature of the climate system when judging how urgent it is to take preventive measures.

Even if present Annex-I Parties' [industrialised countries'] emissions were reduced by 80% by 2100, global emissions would have increased by more than twice the 1990 level, if non-Annex-I Parties [developing countries] emissions were to grow during 1990–2100 according to the non-intervention scenario IS92a.

The reality of this inertia was shown by a simple computation. Scenario IS92a ('Business-as-Usual') implied atmospheric carbon dioxide concentrations in 2010, 2020 and 2030 of 390, 415 and 445 ppmv respectively. Even in the case of the most extreme reduction proposal, IS92c, the atmospheric concentration would continue to increase and reach concentrations of 385, 403 and 423 ppmv respectively. The slow departure from the non-intervention scenario stands out very clearly, and the concentration would still be increasing very significantly in 2030, even if major efforts to reduce the emissions were undertaken. I note in 2007 that in fact the concentration already has reached 380 ppmv and will most likely just about reach 390 ppmv in 2010 as projected in the non-intervention scenario. I continued:

These modest achievements are thus the result of the major inertia of both the climate system *and the socio-economic system of the countries of the world*. If (. . .) stabilization of atmospheric concentrations of greenhouse gases is aimed for, early actions are not noticed quickly because of this inertia but will nevertheless be of major importance in the long term. Similarly a slow start will be difficult to correct for later and will mean losing time. This applies to any country in the world in whatever state of development it finds itself. To take a metaphor that you know well: we are in the situation of turning a big tanker, not in water that can quickly respond to what we do but rather in syrup . . . Also in an operation of this significance, one will of course constantly check on how well one's efforts are bearing fruit, and also how close to the shore we may be, and take further actions as required.

For a long time the IPCC has planned for a closer analysis of how well we are able to assess the importance of natural sources and sinks of greenhouse gases and how these might be modified by human activities. The issue is particularly acute when dealing with the terrestrial ecosystems. In these cases the error margins for the determination of sources and sinks are quite large. It is to be recognized that the net sources or sinks usually are determined as the difference between quite large gross fluxes in both directions between the atmosphere and the terrestrial ecosystems. . . Because of our limited knowledge and lack of observations simplified methods have been proposed . . . for the assessment of sources and sinks by countries. These are . . . very approximate and it is . . . important to analyse their possible short-comings.

This quotation shows that the differences of per capita emissions between developed and developing countries might well become the crucial issue when trying to reach agreements on future emission reductions. The decision in Berlin in 1995 that, to begin with, developing countries would not have to take on quantitative commitments regarding reductions of emissions, expressed an acceptance of this major difference. In the light of the negotiations I therefore judged that Technical Papers 3 and 4 on stabilisation of atmospheric greenhouse gas concentrations ought to be most important.

In retrospect, full recognition that the climate system responds slowly to human interference, that this inertia is hiding what actually is happening and that there are major differences between developed and developing countries had already been given in the SAR and was emphasised in my last presentation to SBSTA in 1997. Nevertheless its political importance was not recognised until almost a decade later.

9.3.2 The USA formulates its strategy on the issue of climate change

It should be clear from this account that the views about a possible future change of the global climate were indeed much divided in the USA. The President and especially the Vice-President, Al Gore, certainly supported the work by the IPCC, but kept a rather low profile. Resistance did not only come from the energy industry, the Senate and Republican congressmen. The third conference of the parties in Kyoto, however, called for a clear statement on the US strategy regarding the global climate change issue. This was given by President Clinton in a talk at the National Geographic Society in October, 1997.[17] The following excerpts from his talk are interesting in the present context:

Today we have a clear responsibility and a golden opportunity to conquer one of the most important challenges of the 21st century – the challenge of climate change – with an environmentally sound and economically strong strategy, to achieve meaningful reductions of greenhouse gases in the United States and throughout the industrialized and the developing world. It is a strategy that, if properly implemented, will create a wealth of new opportunities for entrepreneurs at home, uphold our leadership abroad, and harness the power of free markets to free our planet from an unacceptable risk; a strategy as consistent with out commitment to reject false choices . . .

Greenhouse gas emissions are caused mostly by the inefficient burning of coal and oil for energy . . . The conversion of fuel to energy use is extremely inefficient and could be made much cleaner with existing technologies or those already on the horizon, in ways that will not weaken the economy, but in fact will add to our strength in new businesses and new jobs. If we do this properly, we will not jeopardize our prosperity – we will increase it . . .

First, the United States proposes at Kyoto that we commit to the binding and realistic targets of returning to emissions of 1990 levels between 2008 and 2012. And we should not stop there. We should commit to reduce emissions below 1990 levels in the five-year period thereafter and we must work toward further reductions in the years ahead.

Second, we must embrace flexible mechanisms for meeting these limits. We propose an innovative, joint implementation system that allows a firm in one country to invest in projects that reduces emissions in another country and receive credits for those reductions at home . . .

Third, both industrialized and developing countries must participate in meeting the challenge of climate change. The industrial world must lead, but developing countries must also be engaged . . .

I propose a sweeping plan to provide incentives and lift road blocks . . . to help our companies and our citizens to find new and creative ways of reducing greenhouse gas emissions.

First, we must enact tax cuts and make research and development investments worth up to 5 billion dollars over the next five years – targeted incentives to encourage efficiency and the use of cleaner energy sources.

Second, we must urge companies to take early actions to reduce emissions by ensuring that they receive appropriate credit for showing the way.

Third, we must create a market system for reducing emissions wherever they can be achieved most inexpensively, here or abroad; a system that will draw on our successful experience with acid rain permit trading.

Fourth, we must reinvent how the federal government, the nation's largest energy consumer, buys and uses energy. Through new technology, renewable energy resources, innovative partnerships with private firms and assessment of greenhouse gas emissions from major federal projects, the federal government will play an important role in helping our nation to meet its goal. Today, as a down payment on our million solar roof initiative, I commit the federal government to have 20 000 systems on federal building by 2010.

Fifth, we must unleash competition in the electric industry, to remove outdated regulations and save Americans billions of dollars . . .

Sixth, we must continue to encourage key industry sectors to prepare their own greenhouse gas reduction plans. And we must, along with state and local government, remove barriers to the most energy efficient usage possible . . .

This strategy follows closely the principles and the approach presented in the IPCC SAR, in particular as laid down by Working Group III. The USA had played an important role in the negotiations since 1995 under the leadership of Tim Wirth and Eileen Clauson in the State Department. Dr Watson, cochairman of the IPCC Working Group II, and also assistant to the scientific advisor to President Clinton, was a source of information that was quickly available when needed.

The sceptics had obviously only had a marginal influence on the formulation of the US policy, but it should also be recalled that the Republicans were in majority in the US Senate and that an international treaty, such as a Kyoto Protocol, had to be ratified by the Senate with a two-thirds majority in order to be legally binding. The prospect of achieving this in the near future did not seem very promising and the US government was certainly well aware that the USA might not be able to ratify an agreement reached in Kyoto.

9.4 Increasing industrialisation and globalisation of the world

Even though there was organised resistance to the acceptance of the IPCC analyses of the climate change issue in business circles, some international organisations took the issue more seriously. The Business Council on Sustainable Development was instituted by a group of corporations with the aim of studying the different crucial global issues that might increase in significance over the next few decades, of which climate change was one. I organised some informal contacts with its executive director, Björn Stigson, but the time was not yet ripe for a more organised effort. Instead, industry saw the increasing industrial activities in the developing world, particularly in China, India and Southeast Asia as opportunities for globalisation. Adequate energy supply was of course a prerequisite for rapidly expanding activity in this part of the world. The climate change issue was therefore a rather disturbing element for such future developments.

Nevertheless, it was realised that knowledge and awareness of the global characteristics of climate change was essential in order to meet the future challenges. Concerted efforts were also made by some individuals in industry to be kept well informed about the issue. Throughout my time as IPCC chairman, Brian Flannery, Exxon Research & Engineering Co, in a way served as a 'representative' of the energy industry to the IPCC and was very knowledgeable about global climate change. As a preparation for the upcoming FCCC Conference in Kyoto he wrote quite a detailed status report in November 1997 about the current under-standing of the global climate change issue.[18] Undoubtedly, his analysis was up to date, but his prime focus was on uncertainty. He concluded that there was time to acquire better knowledge before taking action. This conclusion was, however, his subjective judgement and not based on a more penetrating analysis.

I had been in contact with the international industrial organisations dealing with energy, oil and gas and had also been invited to some of their biennial or triennial meetings. As mentioned previously, I attended the meeting of the World Energy Council in Tokyo (1994) and I was on that occasion given an opportunity to present the IPCC analyses. The World Petroleum Congress met again in Beijing in October, 1997. It is interesting to look somewhat more closely at the key presentation by L. R. Raymond, acting president of the Congress (chief executive officer of Exxon) which was presumably partly based on the summary of the climate issue prepared by Flannery.

It is a glossy overview of the necessity of securing adequate energy resources for the development that was about to begin in the Asian–Pacific region. One can, however, easily select at least half a dozen statements regarding global climate change that are not supported by the IPCC analyses and several that are not even supported by Flannery's analysis. It was obviously important to have an outline of future opportunities for negotiations with stakeholders.

9.5 Starting work towards a third assessment

When I was elected as chairman of the IPCC for a second term in 1992, I was given a four-year term in office. Having completed the work for the second assessment at the eleventh IPCC session in Rome, I felt it was desirable that younger people should be involved in the work for the third assessment to be completed by 2000. I also thought that a gradual take-over would be desirable and I was anxious to take the messages from the second assessment to the third conference of the parties to the FCCC in Kyoto in December 1997. I therefore proposed to the IPCC Bureau that the election of a new chairman be held at the twelfth IPCC session scheduled for Mexico City in September 1996 (see IPCC (1996h)) but that he/she should take over at the end of 1997, i.e. after the closing of the FCCC session in Kyoto, and that my own appointment be prolonged accordingly. I also proposed that the IPCC Bureau should be elected at the thirteenth IPCC session scheduled for September–October 1997 in the Maldives. All this was agreed by the Bureau.

The position of chairman was open for nominations and the Bureau appointed a nominating committee under my chairmanship.[19] Dr R. Watson was proposed by the USA. India also announced its interest in nominating a candidate. Other candidates were also put forward. At the formal meeting of the nominating committee, that preceded the plenary session in Mexico City, no Asian candidate was nominated and a unanimously agreed proposal to elect Dr R. Watson was forwarded to the plenary session. Since I had been very careful during my chairmanship not to accept any directives from the Swedish Government in pursuing my role as chairman, I was anxious to secure a similar status for the new chairman. I proposed therefore that such an assurance be given by the US Government for the time during which Watson served as the chairman of the IPCC. I considered this to be essential, because of the necessity for developing countries in particular to feel that a partnership between the scientists of the world is necessary in order for an organisation to gain respect as an independent scientific body. This was agreed at the twelfth session of the IPCC. I do not know to what extent this actually helped Watson during his chairmanship, since he was already perceived to be a defender of the IPCC's independence.

The IPCC tasks after the completion of the SAR had already been discussed at the IPCC plenary sessions in 1994 and 1995, and these constituted the work during the remainder of my term as chairman of the Panel. In November 1996 Watson in turn submitted a proposal for modifications of the organisation of the work towards the Third Assessment Report (TAR) to the IPCC Bureau.

The IPCC met for its thirteenth session in the Maldives on 22 and 25–28 September 1997. The decision to meet in a country consisting of a large number of tropical islands was a recognition of the special threats of flooding to which small island states were exposed. The delegates were reminded by the President of the Republic of the Maldives, Mr Maumoon Abdul Gayoom, of the hardship that had hit the islands during the severe storm and associated inundations in 1987. In fact the sea level had been half a metre above the floor of the hall in which the opening of the session took place.

The meeting was largely concerned with preparing for the major task of next three years, the completion of a TAR, then scheduled for early 2001. The tasks for the three working groups were modified to:

Working Group I will assess the scientific aspects of the climate system and climate change.

Working Group II will assess the scientific, technical, environmental, economic and social aspects of the vulnerability (sensitivity and adaptability) to climate change of, and the negative and positive consequences (impacts) for, ecological systems, socio-economic sectors and human health, with an emphasis on regional, sectoral and cross-sectoral issues.

Working Group III will assess the scientific, technical, environmental, economic and social aspects of the mitigation of climate change, and through a task group (multidisciplinary team) will assess the methodological aspects of crosscutting issues (e.g., equity, discount rates, and decision making frameworks).

The synthesis report will provide a policy-relevant synthesis and integration of the three working group reports.

Robert Watson proposed and it was agreed that the structure of the Bureaux should remain unchanged. However, as often happens when international organisations become established, the fight for positions on executive committees, or in the case of the IPCC, the Bureau, become difficult to resolve. In order to overcome the controversy that emerged in the process of electing members of the Bureau, I finally had to propose that there should be five rather than three vice-chairmen as had been the case during the last five years. I did this without the consultations that would take place under the normal procedure, but I got away with it.[20] Representatives from the developing countries were in the majority, but in the past it had very seldom been necessary to take a vote. The prospects for this continuing to be the case seemed good.

10

The Kyoto Protocol is agreed and a third assessment begun

The role of the IPCC at the third conference of the Climate Convention in Kyoto in December, 1997. Political polarisations increase, while preparations for a third assessment begin.

10.1 Central themes of the Protocol

It soon became obvious that the Kyoto conference was going to be very important in that concerted efforts might be made to reach agreements on binding commitments regarding the future mitigation of climate change. I was reasonably optimistic, since hopefully the delegates were aware of the key conclusions that had been reached by the IPCC in its second assessment, although the controversies during 1996 and 1997 had had an effect on the readiness of parties to commit themselves particularly USA, though countries are in general not easily convinced to do so.

Some 9000 participants gathered in Kyoto for the two weeks that the conference was scheduled to last. Although a number of countries had sent quite big delegations, most of those attending were journalists and representatives of various non-governmental organisations. My summary of the outcome of the conference, as I viewed it at the time, was published in *Science* a few weeks after the end of the conference (see Bolin (1998)). A somewhat modified version of this summary constitutes Sections 10.1–10.4.

Scientific issues were not much discussed in Kyoto. I addressed the Conference on the first day, an IPCC press conference was arranged, and the new IPCC chairman, Dr Robert Watson, addressed the conference during the ministerial segment. The IPCC reports had been used by the delegates during their preparations for the Kyoto Conference as being the most authoritative analysis of climate change (see IPCC, 1996a,b,c,d).

Instead, political and technical issues were in the spotlight. Country delegates positioned themselves with regard to future commitments, that were included in the Protocol which

147

was finally signed after intense negotiations. The Protocol specified different goals for Annex-I (developed) countries and Non-Annex-I (developing) countries. The Protocol would, however, not come into force until 90 days after the date on which 50% parties of the convention ratified it. In addition enough developed countries had to be included to account for at least 55% of their total carbon dioxide emissions in 1990. The USA was at the time responsible for $\sim$38% of these emissions, the EU for 22% and Japan for 8%. This condition meant that the protocol would not come into force until it had been ratified by a number of key developed countries.

10.1.1 Targets and timetables

After long discussions, it was agreed that developing countries would not take on a specific commitment for emission reductions. It was further agreed that a comprehensive approach would be adopted, in that all key greenhouse gases not controlled by the Montreal Protocol on the protection of the ozone layer would be included, i.e. carbon dioxide, methane, nitrous oxide, hydrofluoride (HF_6), perfluorocarbons (PFCs) and sulfurhexafluoride (SF_6). The increase in carbon dioxide accounted then for about 70% of the total increase of radiative forcing. Few measures to decrease emissions of methane and nitrous oxide were available, however. The three other compounds contributed only a small percentage of the total. Nevertheless these was rational decisions.

According to the Protocol, Annex-I parties would take on specific targets, for the reductions of emissions, to be achieved by about 2010. These were given in terms of changes of equivalent CO_2 emissions (Tables 10.1 and 10.2). If parties were able to limit emissions of other greenhouse gases, restrictions on the use of fossil fuels would correspondingly be relaxed. From 1990 to 1995, the EU decreased its carbon dioxide emissions by about 1%, while the other OECD countries together had increased their emissions by about 8% (Table 10.2) and of these Australia, Canada, Japan and the USA had all increased their emissions by 7–9%. Annex-I countries undergoing the transition to a market economy (the Russian Federation and the former Soviet republics), on the other hand, had decreased their emissions by almost 30%. This meant that total emissions of carbon dioxide by all Annex-I parties had already decreased by about 5% from 1990 to 1995.

The targets agreed in Kyoto for Annex-I parties actually added up to a decrease in greenhouse gas emissions to just about 5% below 1990 values (in terms of carbon dioxide equivalents) by 2010. The EU's target was to reduce its emissions from 1% to 8% below 1990 levels, while the other OECD countries were to go from a 7% increase to a 7% reduction, but countries in economic transition were allowed to increase their greatly decreased 1995 emissions by 22–30%,

Table 10.1. *Country commitments for reductions of greenhouse emissions between 1990 and 2010 as a percentage of the equivalent carbon dioxide emissions in 1990; observed changes during the years 1990–1995*

Party	Allowed 1990–2010 (percentage)	Observed 1990–1995
European Union	−8	−1
Austria	−8	−3
Belgium/Luxemburg	−8	+1
Denmark	−8	+18
Finland	−8	+3
France	−8	−4
Germany	−8	−9
Greece	−8	+7
Ireland	−8	−1
Italy	−8	−1
Netherlands	−8	+7
Portugal	−8	+49
Spain	−8	+14
Sweden	−8	+7
UK and N. Ireland	−8	−4
Australia	(−6)	+8
Canada	−6	+9
Iceland	+10	−4
Japan	−6	+8
New Zealand	0	+18
Norway	+1	+9
USA	−7	+7
Bulgaria	−8	n.a.
Croatia	−8	n.a.
Czech Republic	−8	−23
Estonia	−8	n.a.
Hungary	−8	−15
Latvia	−8	n.a.
Poland	−8	n.a.
Romania	−8	n.a.
Russian Federation	0	n.a.
Slovakia	−8	n.a.
Slovenia	−8	n.a.
Ukraine	0	n.a.
Non-Annex-I countries		+25

i.e. to return to 1990 emissions. The Russian Federation and the Ukraine were particularly favoured, and special allowances were also given to Australia, Iceland, New Zealand and Norway. The Protocol thus essentially required a redistribution of emissions between the Annex-I countries.

Table 10.2.   *Emissions of carbon dioxide (Mt C per year)*

Parties	1990	1995
EU	949	936
OECD, except EU	2086	2254
Countries in transition	1311	925
Non-Annex-I countries	1774	2225
Total emissions	6120	6340

From 1990 to 1995, emissions from developing countries had increased by $\sim$25%.[1] Although emissions might not continue to increase at that mean pace, an increase of 4% per year (which was the average growth rate for emissions in OECD countries in the 1960s), would mean that in 2010 Annex-I and Non-Annex-I parties would each contribute $\sim$50% to the total emissions of about 8.3 Gt of carbon, provided that the Annex-I parties complied with the Kyoto agreement. The world population was expected at that time to reach about 7 billion people in 2010 of which about 80% would live in developing countries. Emissions by the developing countries would by then be $\sim$0.74 metric tons of carbon per capita versus about 0.5 metric tons in 1990. On the other hand, the Kyoto agreement meant that per capita emissions in developed countries should decrease from about 3.05 to about 2.90 metric tons of carbon per capita.

10.1.2 Atmospheric carbon dioxide levels

I estimated at the time that the accumulated emissions of carbon dioxide from 1990 to 2010 would amount to about 140 Gt of carbon, which meant that atmospheric carbon dioxide concentrations would increase by about 29 ppmv, i.e. to 382 ppmv with Annex-I countries contributing about 57% and non-Annex I countries about 43%.

This has turned out to be too optimistic a view, primarily because Annex-I countries, except the EU, have increased their emissions significantly (particularly the USA), while countries in economic transition will probably have fully utilised their 'hot air' (i.e. being permitted to return to their emissions in 1990). Developing countries are also expanding their use of fossil fuels more rapidly than foreseen in 1997, particularly China and India. Already in 2007 atmospheric carbon dioxide concentrations are about 380 ppmv and might be close to 390 ppmv in 2010. By 1997, it was already clear that the long residence time of carbon dioxide in the atmosphere meant that even a modest increase in carbon dioxide would be of long-term significance.

10.1.3 Sources and sinks

Atmospheric carbon dioxide concentrations do not change only as a result of burning fossil fuels. The terrestrial biosphere serves as an important source or sink for carbon dioxide as well as for methane and nitrous oxide. Human activities disturb these exchanges, and the Convention stipulates that the parties should report on the effects of anthropogenic interference of this kind. How to account for such terrestrial sinks in the context of national commitments was discussed extensively in Kyoto, particularly with regard to the role of forests.

The IPCC had developed guidelines for establishing a common base for the determination of changes in sources and sinks, but these were not designed on a legal basis for compliance. Changes in carbon inventories in the terrestrial biosphere, including soil, could not be assessed very well. It is difficult to separate anthropogenically introduced changes from natural ones. Various methods had been tried, but the differences between these had not been assessed. The carbon content of soils might change in the opposite direction to changes in above-ground biomass. Regrowth after harvest is influenced by changes in soil conditions and fertilisation. The delayed effects of biogenic processes should be accounted for, but data were lacking and uncertainties large.

Nevertheless the protocol includes the statement that '. . . net changes in greenhouse gas emissions from sources and removal by sinks resulting from direct human-induced land use change and forest activities, limited to afforestation, reforestation and deforestation since 1990, measured as verifiable changes in stocks in each commitment period shall be used to meet the commitments in the Article (No 3) of each Party included in Annex-I . . .' The Protocol also refers to work to be done by the IPCC to resolve this issue before the next conference of the parties.

The discussions about the use of terrestrial sinks were intense because of the potential to compensate for the rather tough reductions of emissions from the use of fossil fuels that had been put on the table by the EU. I tried in this debate to clarify the desirability of being restrictive in considering terrestrial sinks for this purpose. The total amount of carbon in rather rapid exchange between the atmosphere, the upper layers of the oceans and the quickly overturning parts of the terrestrial biosphere (due to the photosynthesis that occurs in the growth of grass, leaves and wood as well as their decay) would increase, since the transfer of this increase to the very large and long-term carbon reservoirs in soils and the deep sea would be slow. The possibility of a later return flow of carbon dioxide to the atmosphere would therefore increase. This argument did not, however, carry much weight in the short-term perspective that prevailed during the negotiations. Referring the matter back to IPCC to be resolved within a year was completely unrealistic. It led, however, to the preparation of the IPCC special report on land-use and use change that was completed a few years later (see IPCC (2000a)).

10.1.4 Tradable emission permits

Although it is important to set targets and timetables, the fundamental problem of climate change cannot be settled that simply. The supply of energy is a fundamental requirement for development. The convention prescribes that '. . . policies and measures to deal with climate change should be cost effective so as to ensure global benefits at the lowest possible cost . . .' Both regulatory measures and economic instruments can be used, but the IPCC had emphasised economic instruments such as emissions trading and carbon taxes that can substantially reduce the costs of achieving a given target. Such policies could raise substantial revenues, and proper distribution of such revenues could dramatically affect the costs of mitigation.

The Kyoto conference was a first step toward the introduction of economical instruments to achieve specific targets. Thus Article 6 of the protocol stipulates: '. . . For the purpose of meeting its commitments under Article 3, any Party included in the Annex I may transfer to, or acquire from, any other such Party emission reduction units resulting from projects aimed at reducing anthropogenic emissions by sources or enhancing anthropogenic removals by sinks of greenhouse gases in any sector of the economy . . .' However, several conditions had to be fulfilled between the parties concerned, i.e. '. . . principles, modalities, rules and guidelines, in particular for verification, reporting and accountability for emissions trading . . .' and would have to be defined by the parties on the convention. This would obviously require time.

The emission limitations and reduction requirements that were agreed in Kyoto as targets for 2010 (Table 10.1) represent, however, an invitation to trade reductions units, particularly between OECD countries and countries in economic transition. The protocol can therefore been seen as an attempt to induce developed countries to find efficient ways of reducing emissions and industry to take appropriate steps. It may also set the stage for gradually creating a global market for emission permits. I felt strongly, however, that care had to be exercised in recognising the major differences between the countries of the world.

The Kyoto conference did not achieve much with regard to limiting the build up of greenhouse gases in the atmosphere. It was clear that if no further steps were taken during the first decade of the twenty-first century, carbon dioxide would increase during the next few decades as it had done in the recent past.

10.2 The interplay of science and politics

The agreement in Kyoto was an attempt to take the first steps towards actually creating a political regime for preventing a human-induced climate change and in that sense was important. The conference accepted without much discussion the view that climate change was on the way and that it would require increasingly demanding efforts in order to limit the emissions of greenhouse gases. It took little notice of the likelihood that there would be a significant climate change and therefore did not much discuss adaptation to future changes. The negotiations focused on assigning fair emission reductions quota to the different countries, and on how to exploit the natural sinks that the terrestrial ecosystems might provide.

It should be recalled that President Clinton had specified that the USA's goal should be not to increase greenhouse gas emissions during the commitment period 2008–2012 above the 1990 level, while the EU had set its goal as decreasing emissions by about 15%, but without really being clear about how to achieve it. This difference in targets was not resolved until US Vice-President Al Gore arrived in Kyoto towards the very end of the negotiations and an agreement could be reached on the commitments for the EU, USA and Japan to be −8%, −7%, and −6% respectively. One may wonder what thoughts were behind this move. It was very clear that it would not be easy for the US Senate (the body that must approve USA participation in international treaties) to make the necessary domestic arrangements. It was obvious from what had happened during 1996 and 1997 that the lobbying by the Climate Council and the Global Climate Coalition and others had to an extent been successful. The resistance in the USA to adopting mandatory commitments meant that in reality it was by no means certain that the Protocol would be ratified soon. As it turned out, President Clinton did not even propose to the Senate that this should be done during his remaining three years in office. The outcome of a vote in the Senate would most certainly be a set-back.

On the other hand, the Russian Federation and the Ukraine were allowed 'hot air', i.e. to increase their emissions back to their levels in 1990. The turmoil that followed the political upheavals in 1989–1991 had led to a substantial decrease in industrial activity and a drastic reduction in the demand for energy.

The protocol also contained a paragraph about 'joint implementation' amongst Annex-1 countries, which allowed the possibility that a country in the need of greater emission allowances than it had been awarded could acquire (buy) these from a country with an excess of such quota. It is to be noted that the inclusion of this possibility in the Protocol had been strongly advocated by the USA based on the success of such an instrument when fighting acid precipitation in the 1980s. Now it might be implemented without US participation.

Paragraph 12 of the Protocol also established a form of cooperation between Annex-I and Non-Annex-I countries (clean development mechanisms, originally proposed by Brazil). Rich countries could assist poor countries by investing in means to reduce their emissions, although they had no formal obligations to do so during the first commitment period, and thereby relax their commitments according to the Kyoto Protocol. The aim was obviously to create a cooperative spirit between industrial and developing countries, which might become important at a later stage when more demanding obligations would most likely be required. The administrative procedures to achieve this were, however, left for further negotiations, and I feared that the cumbersome procedures of the Convention might lead to only slow progress over the years to come.

I suppose that there was unease about the outcome at Kyoto within the US Administration because of the rather tough requirements that the Protocol imposed on industrial nations and not least the USA, but this was not made public. Discussions would of course be resumed at the next conference of the parties, scheduled for Buenos Aires in the autumn of 1998. The lobby organisations were busy and a proposal appeared in the US Senate that US negotiators should not return from the next conference of the parties with commitments for the USA, unless Non-Annex-I countries were also willing to take on binding commitments. This proposal was carried in the US Senate by 95–0 and became a serious hurdle for the negotiations at the convention. This was, however, a breach of the agreements reached in Berlin that the USA had approved in ratifying the Climate Convention.

There was also unease in the Russian Federation, even though its quota were very generous. I exchanged letters with the principal delegate from China to the IPCC, Mr Zou Jingmengh, on the same matter as discussed before. His objections to the final version of the IPCC Technical Paper 4 remained that in the long term the Kyoto Protocol might lead to commitments to reduce emissions in developing countries that would not be acceptable. I had no further communications with China on this subject, but the matter of how developing countries should be brought into a scheme in which all countries would gradually have to take on obligations for reducing greenhouse gas emissions was still unresolved. The Kyoto Protocol was certainly a step forward, although a limited one. Likely future difficulties also became more obvious.

10.3 Opposition to the Kyoto Protocol grows

There was considerable opposition amongst industrialists in the USA to the policy implicitly adopted by the US Administration in agreeing to the final version of the Kyoto Protocol, and this was also the case elsewhere in the world. In April

1998 another initiative was taken by Dr Seitz of the George Marshall Institute in the form of a petition that was circulated widely across the USA with the aim of preventing the ratification of the Kyoto Protocol.[2] The project was supported by an eight-page review of the global warming issue that had been prepared by four researchers at the Oregon Institute of Science and Medicine, and had been signed by about 15 000 scientists. None of the authors had previously published anything dealing with the climate change issue, nor had the review appeared in a peer-reviewed journal. It was, however, printed with a lay-out that was identical to the one used in the *Proceedings of the National Academy of Sciences* (*PNAS*), presumably with the intention of enhancing its credibility amongst readers. The NAS, however, took the extraordinary step of disassociating itself from the initiative of one of its former presidents (i.e. Dr Seitz), expressing its view that the article 'does not reflect the conclusions of expert reports of the Academy'. A closer look at the endless list of names also revealed that few of those that had signed were working in the field of climatology and, as far as I know, none of them was a leading scientist. Actually, a large majority of those that signed were not well informed about the issue at stake.[3] It is of interest to reiterate the issues that were seized upon in the petition to challenge the IPCC views regarding the basic evidence of an ongoing climate change and the reasons for its occurrence because it reflects well the type of information that might preferentially attract the attention of a broad segment of society.

- The source of the recent rise in atmospheric carbon dioxide has not been determined with certainty. In this context reference was made to studies of the very slow variations of atmospheric carbon dioxide concentrations during geological times, which are irrelevant when trying to explain the present rapid increase over a century. There was at the time general agreement amongst scientists in the field that the observed increase during the last 150 years or so was almost exclusively due to human activities.
- The review claimed that the enhanced temperatures in urban areas that result from the heat generated there influenced the determination of the global mean temperature from available surface observations. The careful consideration by the IPCC of the way available data had been used was completely disregarded, as was the fact that temperatures deduced by inverse methods from measurements in deep bore holes far away from densely populated regions showed similar changes during the last century (see IPCC (1996b)). The IPCC had convincingly shown that such errors were insignificant when deducing the global mean surface temperature.
- Satellite data since 1976 showed that the temperature increase in the upper parts of the troposphere seemed to be less than had been derived with the aid of

regular meteorological observations. The authors then *assumed* that the satellite measurements were '. . . the most reliable measurements, and the most relevant to the question of climate change . . .' This was not a valid statement as is commented on in Section 9.3.1 and Section 11.1.1.

- The authors considered that the enhanced rate of photosynthesis that might be expected because of the increasing atmospheric carbon dioxide concentrations would be an overarching benefit for the world. Without closer examination the authors completely disregarded the detailed analyses of available knowledge about the kinds of possible impacts of a significant change of climate on terrestrial systems that had been presented by the IPCC (see IPCC (1996c)).

The challenge was obviously focused on issues that were easily understood by the non-specialist, and it was more difficult for a layman to recognise the inadequacy of the arguments. The analysis was superficial and largely wrong. The scope of the climate issue and available research as summarised and assessed by the IPCC was simply not recognised.

In passing, I wish also to point out an interesting difference between the response to the petition from two representatives of industry, one from the UK and the other from the USA, who were interviewed in the *Wall Street Journal*. Mr Browne, the chief executive of British Petroleum, wanted to find out how the use of fossil fuels could be reduced by introducing a trading system within the BP Corporation of the kind envisaged internationally by the Convention, i.e. to learn about the efficiency of mitigation efforts. Mr Fredrick Palmer, chief executive of the Western Fuels Association, Inc., on the other hand, aimed at '. . . promoting the idea that more carbon dioxide in the air will be good, because it will lead to greater diversity of plants, more abundant crops, and more animals . . .'.[4] I note that in 2007 Mr Browne realised his idea (he had actually presented his vision in 1997), and that British Petroleum by then had '. . . reduced its greenhouse gas emissions by about 10% and reaped huge – and sometimes unexpected – profits, while transforming its traditional oil-field culture and drastically shrinking its so-called carbon footprint . . .'. Although the carbon fertilisation of the terrestrial biosphere due to the increasing carbon dioxide concentration in the atmosphere is a reality, the carbon dioxide uptake from the atmosphere is greatly diminished by the ongoing deforestation.

The experience gained during these years about the interplay between the scientific communities engaged in basic research concerning climate change, stakeholders in society, politicians and the media deserves some further comments. First of all it is obvious that the basic message of global warming is easily understood: the climate of the earth is simply becoming warmer. Almost anyone can therefore join in the discussion. This is different from discussions about many other topics involving new scientific findings with implications for society, e.g. in medical research or the development of modern technology. Scientists from other fields of research become tempted to involve themselves in the climate change debate in a semi-professional manner. They often do not realise the complexity of climate modelling and the difficulty of interpreting available data without serious efforts to comprehend the dynamics of the global climate system, which of course is the prime aim of climate modelling.

The 'overreaction' from society on the environmental issues that have successively emerged as the result of the research during the last few decades was considered in a critical editorial in the *Economist* just after the Kyoto Conference had ended.[5] The target was the environmental scares that have so frequently appeared in the press: 'Forecasters of scarcity and doom are not only invariably wrong; they think that being wrong proves them right.' After having brushed away the pessimistic views of The Club of Rome (Meadows *et al.*, 1972) that were made public in the early 1970s – e.g. diminishing mineral resources, inadequate future food supply, the dangers of air pollution, acid rains – the article turns to some of the principal views concerning global warming. 'Greenhouse warming was originally going to be uncontrolled. Then it was going to be 2.5–4 degrees in a century. Then it became 1.5–3 degrees (according to the United Nations).'

This illustrates above all the fallacies encountered in the transfer of scientific knowledge from a body such as the IPCC to politicians, stakeholders, media and the public at large. The short sentences quoted are wrong and misleading, and of course bewildering for anyone becoming interested or engaged in the issue. The sensitivity of the climate system to external disturbances has essentially remained the same since Manabe in the early 1970s deduced the first range quoted above with the aid of the first three-dimensional climate model. The scares are created when the relevant information is transferred from the scientific community to society at large. Media like hot news. The IPCC has tried to be careful in dealing with matters of this kind. Recommendations are not issued because this necessarily means making implicit societal assumptions, which is not the role of a scientific assessment body. I realise that individual scientists occasionally still do so because of their sincere personal involvement in the issues at stake and also sometimes for political or other reasons. It is further to be noted that some of the early warnings of future environmental catastrophes may seem to have been fear mongering at the time when they were issued. This is now not needed because the measures that have been taken in the meantime have in several cases contributed greatly to improving acute situations.

The rather far-reaching measures taken to avoid serious impacts on forests due to acid precipitation have now markedly diminished the threat of future damage, certainly so in the Scandinavian countries. Lessons have also been learned that are now increasingly being taken advantage of in dealing with the acid rain problem in east and southeast Asia, even though not yet adequately so. The stratospheric ozone depletion issue was not specifically referred to by the *Economist*, but it might have become very serious, if its discovery had not aroused concern amongst people in general and if early action had not been taken. Fortunately, we shall never know to what extent a very serious situation might otherwise have evolved. It should also be stressed that the exaggerations seldom come from the scientific community but instead are often from the media and are underpinned by reports from environmentalists to society at large.

The climate change issue does, however, have additional complicating dimensions. It has indeed taken a long time for it to be generally accepted because of its rather subtle and gradual beginning. But the inertia of global systems also means that if the threats are sensed to be real rather late in the process, it will take a long time to reverse the changes that are on the way. Over the next few generations the projected climate change is next to irreversible. This was gradually realised in wider circles towards the end of the twentieth century, but its possible implications for the future had not then influenced the political debate. There will be reasons to return to this issue later.

Developing countries were also uneasy about the implications of the Kyoto Protocol. Agrawal and Nasrain (1998) pointed out its implicit deficiencies along much the same lines that I had emphasised in my presentations to the IPCC and the SBSTA at my last appearances in 1997. They concluded with a paragraph emphasising that the long term goal had to be achieved by

... aiming for equal per capita emission of greenhouse gases. [This] principle may sound harsh to many industrialized countries. On the other hand, it is a very gracious position for developing countries to take because such a position is only asking for the *future* benefits of the atmosphere to be shared equitably. It is not asking for the factoring in of the past emissions of the industrialized countries, which began with the Industrial Revolution and which have already accumulated in large quantities in the atmosphere.

The article further emphasised that the clean development mechanism is unfair to the developing countries. This is because industrialised countries escape parts of their obligations as defined in the Protocol, i.e. the quotas for reductions of their emissions that they have been allotted, without ensuring that a reduction of global emissions will be achieved. In addition, the article says that the Kyoto treaty legally implies that the assigned amounts of permissible emissions by industrialised countries during the forthcoming commitment period 2008–2012 have gone well beyond being mere targets to be reached. They have been turned into an entitlement by giving developed countries full property rights over these assigned amounts. This provision opened up vast opportunities for 'hot air' to be traded.

Parallel to the IPCC assessments there were rapidly intensified debates of the climate change issue on the Internet that increased further once agreement on the Kyoto Protocol had been reached. I wish to comment on one that emerged in 1998 that illustrates very well the necessity of adhering to the principle of peer review to ascertain scientific credibility.

The IPCC assertion that the increase of the global mean temperature during the latter part of the twentieth century was probably partly due to the increasing concentrations of greenhouse gases and particularly carbon dioxide was challenged. In July 1998 an article by Heinz Hug appeared on the Internet: *The Climate Catastrophe – A Spectroscopic Artefact?*[6] The title of the article of course invites controversy. A most confusing debate began and raged amongst the participants, many of whom were not specialists in the field of radiative transfer in gases with absorption in the infrared (IR) part of the spectrum (e.g. carbon dioxide, water vapour etc.). Two fundamental mistakes had been made in the analysis that served as the basis for the conclusions drawn. (1) A laboratory experiment had been set up to determine the absorption spectrum of carbon dioxide in the IR range that was technically inadequate and also unnecessary, because such measurements had been done carefully long ago. (2) The use of this basic information for the study of energy transfer in an atmosphere containing carbon dioxide with a concentration decreasing upwards, and in the presence of other greenhouse gases with partly overlapping spectra requires a careful integration step by step up through the atmosphere. This was not done. The discussion went on for several months and then disappeared quickly. The matter had been analysed by climate modellers decades before. A more thorough presentation had, however, not been included in the IPCC reports, but references merely made to the standard literature in the field in the 1980s and earlier. Strangely a similar debate appeared amongst some sceptics in Sweden as late as in 2006 and was initiated by researchers at the Royal Institute of Technology (KTH) in Stockholm, I am sure that there have been similar debates on the Internet before and there will be others in the future causing unnecessary doubts about the basic achievements by the leading researchers in the field. Criticism is welcome but should be pursued through the peer-reviewed literature and then, if appropriate, become part of mainstream scientific knowledge.

10.4 How to settle disagreements on the interpretation of the Kyoto Protocol

The agreement on the Kyoto Protocol was certainly an important achievement, but in retrospect at the time of its adoption it was already politically rather unrealistic. It was very succinct with regard to the quotas that were allotted to countries, but, for example, an interpretation was needed regarding paragraph 3, which specified the way in which countries could use terrestrial sinks in order to fulfil their obligations. The rather vague language in several of the paragraphs had to be clarified before the ways and means of implementing the decisions in Kyoto were agreed. In addition each country wished primarily to limit its own obligations, which clearly was a sign of them not yet recognising the global issue adequately. The fourth conferences of the parties in 1998 in Buenos Aires was intended to be a step towards a more carefully prepared conference of the parties (the sixth) planned for Amsterdam in November 2000.

The more precise implications of the Kyoto agreement and how to turn its articles into practical actions became the key items for the discussions in Amsterdam. The IPCC assessment had shown that the terrestrial ecosystems, including forests and soils, serve as significant natural sinks for carbon dioxide, and the crucial question emerged: how were the means to exploit this feature of the natural carbon cycle to be evaluated quantitatively in order to fulfil the commitments as agreed in the Protocol? Based on the IPCC analyses I had already warned in Kyoto that quantitative estimates of such sinks would be very uncertain and might well supply loopholes for countries that were unable to fulfil their commitments. This fundamental aspect of the global carbon cycle was not adequately recognised in the negotiations in spite of the publication of a short and extremely relevant article in May 1998 by the IGBP, in which the characteristics of the terrestrial system that are important in this context were emphasised[7]:

According to the Kyoto Protocol, Annex I countries can reduce emissions by limiting fossil fuel consumption or by increasing net carbon sequestration in terrestrial carbon sinks. The inclusion of carbon sources and sinks in a legally binding emissions reduction framework is significant. However, it creates a number of problems that, if not corrected, will seriously limit the protocol's effectiveness . . . About 50% of the initial uptake of carbon through photosynthesis, Gross Primary Production, (GPP, globally ~120 Gt/yr) is used by plants for growth and maintenance. The remaining carbon [after respiration] is Net Primary Production (NPP, globally ~60 Gt/yr). Part of this is shed as litter and enters the soil, where it decomposes, releasing nutrients to the soil and carbon dioxide to the atmosphere. The remaining carbon after these emissions is Net Ecosystem Production (NEP, globally ~10 Gt/yr). Much of this is lost by non-respiratory processes such as fire, insect damage, and harvest. The remaining carbon is called Net Biome Production (NBP, globally ~2 Gt/yr) being a small fraction of the initial uptake of carbon dioxide from the atmosphere and can be positive or negative; at equilibrium it would be zero. NBP is the critical parameter to consider for long-term (decadal) carbon storage.

It is obviously difficult to determine the NBP accurately because it is a small fraction of the NPP (at most merely about 5%, or <3 Gt C per year). In addition its spatial distribution is very patchy. The global deforestation and changing land use at the turn of the century had been estimated with the aid of global decadal inventories to be about 1.6 Gt C per year. The observations needed to assess adequately the carbon sources and sinks in the exchange between the atmosphere and the soil at local and regional levels are not easily extracted from available data so the final estimate of the net transfer is very uncertain.

Just a few weeks before the Buenos Aires meeting in 1998, Fan *et al.* (1998) published an assessment of the exchange of carbon dioxide between the atmosphere and the terrestrial systems, using a so-called inverse analysis based on observations of the variations of the carbon dioxide concentration distribution around the earth during the year. It showed that North America was a sink for carbon dioxide during the last decade of the twentieth century, contrary to what was found to be the case for the Eurasian continent. The analysis was, however, severely criticised because of the limited and inadequate data base that had been used, which did not allow reliable conclusions to be made. The paper influenced the final political negotiation in Amsterdam two years later, in spite of the fact that the study had not yet been reviewed by the IPCC.

It is difficult to determine accurately the uptake of carbon dioxide by or its release from the terrestrial systems and therefore also difficult to arrange satisfactory accounting measures in order to ensure that the reporting would not cause troubles later. Based on Fan *et al.* (1998), the USA and Canada wanted to exploit the carbon dioxide sinks that forests provide, while the EU strongly objected to such an action. However, terrestrial sinks do not adequately compensate for continued emissions from fossil fuel burning into the atmosphere. As has been emphasised before, the total stock of carbon circulating between the rather rapidly exchanging natural pools, i.e. (1) the atmosphere, (2) the terrestrial pool of leaves, grass, and the uppermost layers of the soils (including litter) and (3) the surface layers of the oceans, would still be increasing, because the transfer of the amounts of carbon dioxide emitted due to fossil fuel burning into the deep layers of the oceans and the deeper layers of soil is very slow. Return flow to the atmosphere might therefore occur in the future. Attempts were made in the negotiations in Amsterdam to resolve these issues by accepting just an agreed percentage of the reported uptake by the terrestrial biosphere, but it was too late in the negotiations to reach a compromise.

Instead it was agreed to adjourn the meeting and call for a resumption of this sixth conference of the parties in mid-2001. This, however, ultimately meant that the whole process of establishing a regime for combating climate change was seriously delayed because George W. Bush was elected US President late in 2000. However, an agreement on the Kyoto Protocol in Amsterdam would not have eliminated the difficulties. The US Senate at the time would not have agreed to this international treaty.

The agreement on the Kyoto Protocol was a political compromise. National and economic interests were brought into focus and the governments of several developed countries asked the question: is it really possible to reach the specified goals without a substantial reduction in the rate of economic growth? The analyses behind this question were, however, mostly narrow, inadequate and, in particular, short-sighted. It was, however, difficult to attract political attention and obtain agreements for measures that lasted more than a decade. The first commitment period in the Kyoto Protocol, i.e. 2008–2012, still seemed far away. The view expressed in the SAR and put forward by the IPCC at the Kyoto Conference that the proposed measures would only be the first steps towards a stabilisation of greenhouse gas concentrations in the atmosphere, was not adequately recognised. Short-sighted economics as well as a lack of conviction that the climate change issue really might become a serious threat dominated the minds of the decision-makers. The agreement by the US Senate in a 95–0 vote not to accept a compulsory requirement for the USA unless developing countries also took on quantified obligations, became a serious obstacle to progress. This vote was instigated by the key industrial lobby organisations that I have previously referred to and that I knew well. It was a clear protest against the agreement reached in 1995 at the first conference of the parties in Berlin. This limited the possibility for the US Administration to act as a steward for the world as formulated by President George Bush, senior, in his address to the IPCC in 1989. That role for the USA had largely been forgotten ten years later.

11

A decade of hesitance and slow progress

While the political process of combating climate change almost came to a standstill, the global warming continued, but the assessments of new scientific findings progressed.

11.1 Work towards the IPCC Third Assessment Report

The eleventh plenary session of the IPCC in the Maldives in October 1997 and the third conference of the parties to the Climate Convention in Kyoto in December that same year were my last engagements as chairman of the IPCC. Since then I have followed the IPCC work attentively but at a distance, have taken part in the work on IPCC's *Special Report on Land Use, Land-Use Change, and Forestry*, and been one of the review editors of *The Scientific Basis of the Third IPCC Assessment Report* published in 2001.[1] Obviously, I have not been in a position to follow the internal work and controversies as closely as during my time as chairman of the IPCC. Nonetheless the following observations might be of interest.

First, five major special reports were completed during 1998–2001.

- *Analyses of the regional impacts of climate change (IPCC, 1998).* In addition to this work by the IPCC, a large number of other analyses have of course been carried out at national level. This is not, however, the place to expand on this essential work.
- *Emissions of greenhouse gases to the atmosphere from aviation (IPCC, 1999).* The IPCC began work for this report in the mid-1990s and this marked the start of more elaborate cooperation between the IPCC and stakeholders, on this occasion through the International Civil Aviation Organisation, which turned out to be very useful for both parties. Throughout my time as IPCC chairman I maintained the view that wherever possible collaboration with interested parties was essential in the long term. The prime issue in this particular context

163

was an assessment of the current and likely future contributions of greenhouse gases emissions (primarily carbon dioxide and water vapour) from the rapidly increasing civil aviation industry to the radiative forcing in general. This contribution was estimated to be about 0.04 W m^{-2}, merely about 1½% of all human emissions but it was believed it might roughly triple during the next 50 years. Air traffic also creates contrails, which are estimated to cover about 0.1% of the earth's surface. They tend to warm the earth like thin high level clouds, and at present contribute about 0.06 W m^{-2} to the enhanced greenhouse effect. In addition, high-level air traffic enhances the formation of wide-spread cirrus clouds, the significance of which is more difficult to assess, but might be a more important contributer to the radiative budget of the atmosphere. These effects are obviously not of major significance, but must not be ignored.

- *Land-use, land-use change and forestry (IPCC, 2000a).* The work on this report was intended to serve as a basis for the development of the rules of procedure to be used in accounting for sources and sinks in the implementation of the Kyoto Protocol and has been dealt with briefly in Section 10.1 The outcome of the analyses demonstrated the difficulties that might arise. It is difficult to define 'a forest' because of the many different types that are found around the world. I felt that these sorts of problems could only worsen when the rules of procedure were applied in a political context, and that it would be difficult to ensure that the procedures were fair and accurate enough for their purpose. I had suspected in Kyoto that this might prove problematic and this report seemed to confirm that suspicion.

- *Emission scenarios (IPCC, 2000b).* The *Special Report on Emission Scenarios, SRES,* presents about 40 scenarios illustrating different socio-economic development paths for the world, aimed at embracing the likely total range of greenhouse gas emissions scenarios during the twenty-first century. Four of these were elaborated in terms of so-called storylines:

  > *The A1 storyline and scenario family* describes a future world of very rapid economic growth, low population growth (maximum about 9 billion around 2050, then declining to about 7 billion in 2100) and the rapid introduction of new and more efficient technologies. Major underlying themes are convergence between regions, capacity building, and increased cultural and social interactions, with a substantial reduction in regional differences in per capita income.

  > *The A2 storyline and scenario family* describes a very heterogeneous world. The underlying theme is self-reliance and preservation of local identities. Fertility patterns across regions converge very slowly, which results in high population growth. Economic development is primarily regionally oriented and per capita economic growth and technological change is more fragmented and slower than in other storylines.

The B1 storyline and scenario family describes a convergent world, with the same population growth as in A1 but with rapid change in economic structures towards a service and information economy, with reduction in material intensity, and the introduction of clean and resource-efficient technologies. The emphasis is on global solutions to economic, social and environmental sustainability, including improved equity, but without specific climate initiatives.

The B2 storyline and scenario family describes a world in which the emphasis is on local solutions to economic, social and environmental sustainability. It is a world with moderate population growth, intermediate levels of economic development, and less rapid and more diverse technological change than in the B1 and A1 storylines. While the scenario is also oriented towards environmental protection and social equity, it focuses on local and regional levels.

Work on the scenarios began in 1996 and required great care since it was important not to be caught once more in a critical debate about the trustworthiness of such analyses. The very many assumptions made indeed justified that great care be exercised in the interpretation of the results. Nevertheless, it was essential that a 'common set' of scenarios was available for use by the climate modellers and that the assumptions made were accurately described. This new set of scenarios was used extensively during the following years.

• *Methodological and technological issues in technology transfer (IPCC, 2000c).* These issues will not be dealt with here but are obviously important for industry as the Convention agreements are implemented.

It had been agreed that the TAR should appear early in 2001. Work began in 1998 and was pursued in parallel with the completion of the special reports. During 1998–2001 nine major reports were thus completed that together comprise more than 5000 pages and 25 000 references to papers in the relevant scientific and technical literature. This obviously meant a very heavy burden for the numerous scientists and experts that were engaged in the work and in particular for the technical support units attached to the three working groups. It occasionally also meant that key scientists became hesitant to become involved, since it meant that much less time would be available for their own research.

The IPCC recognised the difficulty for any individual or institution to grasp fully what was brought together in this way and again prepared a synthesis report of the third assessment, which required much more work than that for the first and second reports (IPCC, 2001d). It was based on altogether nine questions that had been formulated on the basis of requests from country delegates to the Climate Convention. The resulting 145 pages including a summary for policy makers (34 pp) and numerous illustrations provide a detailed guide to the knowledge base available at the turn of the century, but did it help to convey the key message widely?

It is still difficult to extract more precise information on matters that will concern people as individuals and an obvious sign of this difficulty is the way the issue of extreme events is being dealt with in the public debate. Heat waves and droughts, excessive precipitation and flooding, and devastating storms and gales are considered to be major threats that are already hitting us but it is difficult to extract from the synthesis report how to deal with questions such as: to what extent, where, when and how badly will such changes of the climate come upon us? Extreme weather events have always occurred and the basic question is obviously: to what extent will society be more adversely affected by them? What will the most trustworthy scenarios for the future in this regard look like? In order to be able to take action it is necessary to provide the general public with information about the measures to be taken, with regard to both mitigation and adaptation. The slow progress in the almost 10 years since the agreement on a Kyoto Protocol may partly be due to this difficulty, although obviously other immediate concerns in world politics have also played an important role, particularly the wave of terrorism that swept the globe.

The TAR was completed early in 2001. The following brief account of its content and conclusions is necessarily subjective in the selection of topics. I have focused on the parts that are, as far as I can judge, central to developing a strategy for addressing the climate change issues. This should be aimed at slowing and ultimately stopping the climate change, but, as it will not be possible to achieve this in less than 50 years, it will be necessary to include plans for adaptation. The adaptation required will depend on the particular circumstances of each country. It will not be analysed further here; focus will rather be on the truly global issue of mitigation.

The extra five years of observations since 1995 and the new analyses that were available by the turn of the century clarified a number of controversial issues. However, the complexity of the climate change issue also became apparent and this sometimes sharpened the controversies.

11.1.1 The Working Group I contribution to the TAR: scientific basis; summary for policy makers, adopted 17–20 January 2001 in Shanghai (see IPCC (2001a))

A selected set of key conclusions as formulated in the summary for policy makers read as follows (the reader is referred to the summary for policy makers as a whole and the synthesis report for a more complete analysis and a number of informative illustrations).

An increasing body of observations give a collective picture of a warming world and other changes of the climate system

'The global average surface temperature has increased since 1861. Over the twentieth century the increase has been $0.6 \pm 0.2\,°C$ . . . These numbers take into account various adjustments, including urban heat island effects. The record shows a great deal of variability; for example, most of the warming occurred during the twentieth century during two periods, 1910–1945 and 1976–2000 . . .'

The global mean temperature has continued to increase slowly since the turn of the century and the increase was estimated to be almost $0.8 \pm 0.2\,°C$ by 2007; see Section 12.2 and Figure 12.1.

New analyses of proxy data from the Northern Hemisphere indicate that the increase in temperature during the twentieth century is likely to have been the largest of any century during the past 1000 years. It is also likely that, in the Northern Hemisphere, the 1990s was the warmest decade and 1998 the warmest year.[2]

This conclusion that was drawn in a study by Mann *et al.* and was partly supported by Crowley.[3] It was, however, severely criticised a few years later by McIntyre and McKitrick with the conclusion that the particular "hockey stick" shape derived by Mann *et al.* proxy construction is primarily an artefact of poor data handling, obsolete data and incorrect calculation of principal components . . .'[4] Their analysis is hardly more reliable than the one published by Mann *et al.* It is also noteworthy that their criticism focused on the credibility of the Mann analysis, rather than the uncertainty of any analysis because of the very limited amounts of data available which makes any interpretation rather uncertain. Other reconstructions have since then supported Mann's conclusions, see Figure 11.1.[5]

The IPCC continues with a number of succinct conclusions:

We find no evidence for any earlier periods in the last two millennia with warmer conditions than the post-1990 period – in general agreement with previous similar studies. The main implications of our study, however, are that natural multi-centennial climate variability may be larger than commonly thought.

Since the late 1950s the overall global temperature increases in the lowest 8 kilometres of the atmosphere and in surface temperature have been similar at $0.1\,°C$ per decade.[6]

There are also observations of soil temperature changes down to considerable depths that can be used for the determinations of the temperature changes at the earth's surface during the last 100 years that support the regular meteorological observations.[7]

Satellite data show that it is very likely there have been decreases of about 10% in the extent of snow cover since the late 1960s, and ground observations show that it is very likely there has been a reduction of about two weeks in annual duration of lake and river ice cover in the mid and high latitudes of the Northern Hemisphere over the twentieth century.

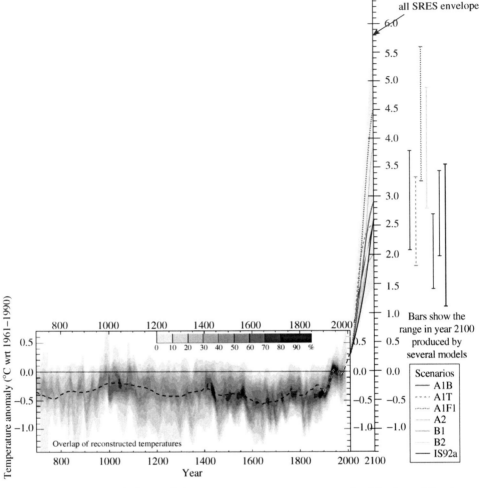

Figure 11.1 Variations of the average temperature over the Northern Hemisphere during the last 1300 years. Instrumental records have been available since the latter part of the nineteenth century and earlier variations have been reconstructed using a variety of indirect means; the uncertainty ranges of these analyses are also shown (IPCC, 2001a, 2007a). A range of changes of global mean temperature due to human emissions during the twenty-first century based on the SRES emission scenarios, shows an uncertainty by 2100 of between 1.5 and 5.8 °C.

Northern Hemisphere spring and summer sea ice extent has decreased about 10 to 15% since the 1950s. It is likely that there has been about a 40% decline in the Arctic sea-ice thickness during late summer and early autumn in recent decades.

Tide gauge data show that global average sea level rose between 0.1 and 0.2 metres during the twentieth century.

It is very likely that precipitation has increased by 0.5 to 1% per decade in the twentieth century over most mid and high latitudes of the Northern Hemisphere continents and it is likely that rainfall has increased by 0.2 to 0.3% per decade over the tropical land areas. It is also likely that rainfall has decreased over much of the Northern Hemisphere subtropical land areas during the twentieth century by about 0.3% per decade.

In the mid and high latitudes of the Northern Hemisphere over the latter half of the twentieth century, it is likely that there has been 2 to 4% increase in the frequency of heavy precipitation events.

It is likely that there has been a 2% decrease of cloud cover over mid- to high-latitude land during the 20th century.

On the other hand, some important aspects of climate appear not to have changed:

A few areas of the globe have not warmed in recent decades mainly over some parts of the Southern Hemisphere oceans and parts of Antarctica.

No significant trends of Antarctic sea ice extent are apparent since 1978, the period of reliable satellite measurement.

Changes in tropical and extra-tropical storm intensity and frequency are dominated by inter-decadal and multi-decadal variations, with no significant trends evident over the twentieth century. Conflicting analyses make it difficult to draw definitive conclusions about changes in storm activity, especially in the extra-tropics.

No systematic changes in the frequency of tornados, thunder days, or hail events are evident in the limited areas that have been analysed.

I wish to emphasise that the points in the last four quotations have not been taken note of adequately by the media during the five years since the TAR was released. This very fact deserves more attention and will be addressed later.

Emissions of greenhouse gases and aerosols due to human activities continue to alter the atmosphere in ways that are expected to affect the climate

The atmospheric concentration of carbon dioxide, CO_2, has increased by 31% since 1750 [to 368 ppmv]. The present concentration has not been exceeded during the last 420 000 years and possibly not during the past 20 million years.

About three-quarters of the anthropogenic emissions of CO_2 to the atmosphere during the past 20 years are due to fossil fuel burning. The rest is predominantly due to land use change, especially deforestation.

This conclusion was later further tested by comparisons between model projections and data for about the last 50 years with the conclusion that no other hypotheses about a causal relation between suggested physical mechanisms and observed temperature changes have (as yet) turned out to be significant.

RADIATIVE FORCING COMPONENTS

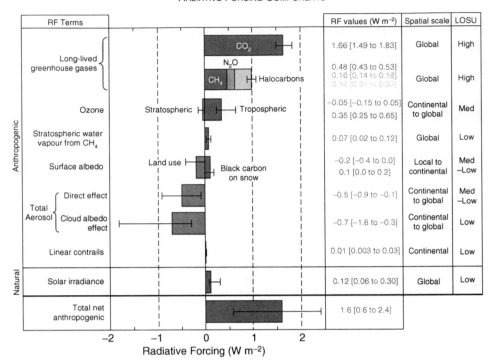

Figure 11.2   The global mean radiative forcing of the climate system for the year 2005, relative to 1750 due to human emissions of greenhouse gases and aerosols. Alterations of surface reflectance due to changes in land use and variations of the output of the sun are also shown. Forcing due to episodic volcanic events that lead to negative radiative forcing lasting only for a few years is not shown (IPPC, 2007a).

Currently the ocean and the land together are taking up about half of the anthropogenic CO_2 emissions. On land, the uptake of anthropogenic CO_2 very likely exceeded the release by deforestation during the 1990s.

The atmospheric concentration of methane has increased by 150% since 1750 and continues to rise.

The radiative forcing due to the increases of the well mixed greenhouse gases (carbon dioxide, methane, and other anthropogenic greenhouse gases) from 1750 to 2005 and their uncertainties are shown. Anthropogenic aerosols are short-lived and mostly produce negative forcing. The major sources are fossil fuel and biomass burning (see Figure 11.2, which has been extracted from the Fourth Assessment Report to provide an up-to-date view).

In addition to their direct radiative forcing, aerosols have an indirect radiative forcing through their effects on clouds. It is mostly negative although of uncertain magnitude.

Natural factors have made small contributions to changes of radiative forcing over the past century.

The *observed changes* of the climate have been analysed using global climate models in order to determine to what extent they might be caused by human activities (see also Stott *et al.* (2001)).

Simulations that include estimates of natural and anthropogenic forcing reproduce reasonably well observed large-scale (global mean) changes in surface temperature over the twentieth century. However, contributions from some additional processes and forcing may not have been included in the models. Nevertheless, the large-scale consistency between models and observations can be used to provide an independent check on projected warming rates over the next few decades under a given emissions scenario.

The warming during the past 100 years is very unlikely to be due to internal variability alone.

Detection and attribution studies consistently find evidence for an anthropogenic signal in the climate record of the last 35 to 50 years.

Simulations of the response to natural forcing alone (i.e. the response to variability in solar radiation and volcanic eruptions) do not explain the warming in the second half of the twentieth century. However, they indicate that the natural forcing may have contributed to the observed warming in the first half of the twentieth century.

These conclusions should be compared with the comment made in the FAR: 'The size of this warming is broadly consistent with predictions of climate models, but is also of the same magnitude as natural variability. Thus the observed increase could largely be due to this natural variability; alternatively this variability and other human factors could have offset a still larger human-induced greenhouse warming.'

Human influences will continue to change atmospheric composition throughout the twenty-first century

Emissions of CO_2 due to fossil fuel burning are virtually certain to be the dominant influence on the trends in atmospheric CO_2 concentration during the twenty-first century.

As the CO_2 concentration of the atmosphere increases, ocean and land will take up a decreasing fraction of anthropogenic emissions.

By 2100, carbon cycle models project atmospheric CO_2 concentrations of 540 to 970 ppmv for the *SERS* scenarios.

Model calculations of the concentration of non-CO_2 greenhouse gases by 2100 vary considerably across the *SRES* scenarios.

Global average temperature and sea level are projected to rise under all IPCC SRES scenarios

The globally averaged surface temperature is projected to increase by 1.4–5.8 °C over the period 1990–2000. These results are for the full range of *SRES* scenarios and based on a number of climate models (see Figure 11.1).

Based on recent global models simulations, it is very likely that nearly all land areas will warm more rapidly than the global average, particularly those at northern high latitudes in the cold season. Most notable of these is the warming in the northern regions of North America and northern and central Asia, which exceeds global mean in each model by more than 40%. In contrast, the warming is less than the global mean change in south and southeast Asia in summer and in southern South America in winter.

Global average water vapour concentration and precipitation are projected to increase during the twenty-first century.

Extreme events

Projected changes during the twenty-first century indicate that it is *very likely* that there will be

> higher maximum temperatures and more hot days over nearly all land areas,
> higher minimum temperatures and fewer cold days over nearly all land areas,
> reduced diurnal temperature range over most land areas,
> more intense precipitation events over many areas.

it is *likely* that there will be

> increased summer, mid-continental drying and associated risk of drought,
> increase in tropical cyclone peak wind intensities,
> increase in tropical cyclone mean and peak precipitation intensities in some areas.

Some special features of the expected changes

Even with little or no change in El Niño, the amplitude of global warming is likely to increase the risk of droughts and floods.

It is likely that warming associated with increasing greenhouse gas concentration will cause an increase of Asian summer monsoon precipitation variability.

Most models show weakening of the ocean thermocline circulation,[8] which leads to a reduction of the [ocean] heat transport into high latitudes of the Northern Hemisphere.

Northern Hemisphere snow cover and sea-ice extent are projected to decrease further. Glaciers and ice sheets are projected to continue their widespread retreat during the twenty-first century.

The Antarctic ice sheet is likely to gain mass because of greater precipitation, while the Greenland ice sheet is likely to lose mass because the increase in run off will exceed the precipitation increase.

Global mean sea level is projected to rise by 0.09–0.88 metres between 1990 and 2100 for the full range of SRES scenarios, primarily due to thermal expansion of sea water and loss of mass from glaciers and ice caps.

Anthropogenic climate change will persist for many centuries Emissions of long-lived greenhouse gases have a lasting effect on atmospheric composition, radiative forcing and climate.

Global mean surface temperature increases and rising sea level from thermal expansion of the ocean are projected to continue for hundreds of years after stabilisation of greenhouse gas concentrations.

Ice sheets will continue to react to climate warming and contribute to sea level rise for thousands of years after climate has stabilised.

I have chosen to quote the summary for policy makers quite extensively in order to bring home to the reader the flavour of its basic character of being factual and balanced, but still in some regards qualitative. It is of course necessary to penetrate deeper into the individual chapters in order to check on the validity of the conclusions that have been drawn. When this has been done, there have seldom been disagreements between the conclusion in the bulk reports and the summary for policy makers (National Research Council, 2001). Admittedly, the uncertainties that are associated with several of the conclusions drawn may not always have been specifically stressed in the summaries for policy makers. Few politicians and stakeholders check the internal consistency in the IPCC reports and confidence in the IPCC process is therefore of fundamental importance. This can only be achieved if the scientific community at large agrees with the conclusions drawn by the IPCC.

11.1.2 Working Group II contribution to the TAR, impact, adaptation and vulnerability: summary for policy makers adopted 13–16 Feb. 2001 in Geneva (IPCC, 2001b)

The task for Working Group II was in principle quite different from that pursued by Working Group I, in that impact, adaptation and vulnerability concern people and countries that are affected by climate change and that respond differently because of major differences in their geographical and environmental setting and socio-economic status. This is also reflected in the large number of scientists that became engaged in the assessment process and the numerous papers and reports that were needed as a basis for the analyses. The report was written by about 180 convening lead authors and lead authors and about 240 contributing authors. It was reviewed by about 440 government and expert reviewers and 33 review editors oversaw the review process. But most of these were primarily concerned with the effects in their own countries or regions of the world.

Many parts of the report make interesting reading, but cannot easily be summarised in global terms. The eight chapters of regional analyses need to be evaluated in terms of the socio-economic characteristics by individual countries in their respective regions in order to permit more specific conclusions to be drawn. The assessment might thus serve as a starting point for national initiatives.

Analyses of what a future climate change might bring with it necessarily also require better understanding of present vulnerabilities which actually are poorly known in many countries. Analyses of the impacts of a human-induced climate change might then also on many occasions reveal that insufficient measures have been taken to provide adequate security against extreme events that are part of the *prevailing* climate, which in itself might justify taking protective measures now. There are many examples that support this statement.

The large differences between regions are exemplified by quoting some of the conclusions drawn for *Africa* and *Europe* respectively in the summary for policy makers:

Adaptive capacity of human systems in *Africa* is low due to the lack of economic resources and technology, and vulnerability is high as a result of heavy reliance on rain-fed agriculture, frequent droughts and floods, and poverty.

Grain yields are projected to decrease for many scenarios, diminishing food security, particularly in small food importing countries.

Major rivers of Africa are highly sensitive to climate variations; average runoff and water availability would decrease in Mediterranean and southern countries of Africa.

Extension of ranges of infectious disease vectors would adversely affect human health.

Desertification would be exacerbated by reduction in average annual rainfall, runoff and soil moisture especially in southern, North, and West Africa.

Increase in droughts, floods, and other extreme events would add to stresses on water resources, food security, human health, and infrastructures, and would constrain development.

Significant extinctions of plant and animal species are projected and would impact on rural livelihood, tourism and genetic resources.

Coastal settlement in, for example, the Gulf of Guinea, Senegal, Gambia, Egypt and southern and southeastern African coasts would be adversely impacted by sea level rise through inundations and coastal erosion.

Adaptive capacity is generally high in *Europe* for human systems; southern Europe and European Arctic are more vulnerable than other parts of Europe.

Summer runoff, water availability and soil moisture are likely to decrease in southern Europe, and would widen the difference between the north and drought-prone south.

Half of Alpine glaciers and large permafrost regions could disappear by the end of the twenty-first century.

River flood hazards will increase across much of Europe; in coastal areas, the risk of flooding, erosion, and wetland loss will increase substantially with implications for human settlement, industry, tourism, agriculture, and coastal natural habitats.

There will be some broadly positive effects on agriculture in northern Europe; productivity will decrease in southern and Eastern Europe.

Upward and northward shift of biotic zones will take place. Loss of important habitats (wetlands, tundra, and isolated habitats) would threaten some species.

Higher temperatures and heat waves might change traditional summer tourist destinations and less reliable snow conditions may impact adversely on winter tourism.

The impacts of climate change as well as vulnerability and adaptive capacity of people and countries depend on

- the slow changes of average climatic conditions that often are obscured because of weather variability,
- the occurrence of extreme events.

However, the seriousness, timing and spatial distribution of extreme events ('hot spots') cannot be well foreseen. Also, Working Group I was cautious when describing expected future changes in their occurrence, since scientific evidence is still meagre.

This dilemma is found throughout the report. It probably also had an impact on the evaluation of the report. Environmentalists took note of the large number of possible impacts and often assumed high vulnerability, while sceptics found the conclusions in the report were not always adequately supported by observations. Of course it is not difficult to find extreme events in the past that have had serious impacts but that did not occur because of a climate change, and single events today cannot easily be associated with the global warming that is on its way. Statistical analyses are needed to settle *whether there are systematic trends towards other weather regimes or not and to what extent they might be related to the ongoing global warming.* Quantitative studies of this kind are still few, but the observational evidence gathered in the report indicates qualitatively that the impact of global warming might well be one reason, and a key one, for the observed changes. This is a most crucial issue in the process of developing mitigation strategies. We shall also have reasons to return to this discussion in summarising the outcome of the assessment by Working Group III.

11.1.3 Working Group III contribution to the TAR, mitigation: summary for policy makers adopted in Accra Ghana, 28 February to 3 March 2001 (IPCC, 2001c).

Global climate change is a truly global problem. Mitigation will therefore necessarily require international cooperation and recognition that developed and developing countries must work in concert to find ways and means to slow down and to stop ongoing changes. I will focus on this aspect of climate change in the concluding chapters.

Most appropriately the Working Group III report opens with a chapter on 'Setting the stage: climate change and sustainable development'. The key factor that is bringing about a possibly serious change of the global climate is the world's rapidly increasing demand for energy, since it is a prime prerequisite for industrialisation and many countries in the world still have an inadequate industrial infrastructure. Demand for energy will also increase because of the increase of world population by about 1% annually, although this rate is decreasing slowly.

At present some 80% of the world energy supply comes from the burning of fossils fuels and this is the prime cause for the emerging climate change. I quote from the technical summary of the Working Group III report:

During the period of intense industrialisation from 1860 to 1997, an estimated 13 000 EJ of fossil fuels were burned,[9] releasing 280 Gt C into the atmosphere, which along with the land-use change has raised atmospheric concentrations of carbon dioxide by 30%. Estimated natural gas resources are comparable with those for oil, being approximately 35 000 EJ. The coal resource base is approximately four times as large. Methane clathrates (not included in the *resource* base) are estimated to be approximately 780 000 EJ. Estimated fossil fuel *reserves* contain 1500 Gt C, being more than five times the carbon already released, and if estimated resources are added, there is at least a total of 5000 Gt C in the ground. On top of that there are additional quantities with unknown certainty of occurrence, referred to as 'additional occurrences'. Examples of unconventional fossil fuel resources are tar sands and shale oils, geo-pressured gas, and gas in aquifers.[10]

The scenarios modelled by SRES without any specific greenhouse gas emission policies foresee cumulative release ranging from approximately 1000 Gt C to 2100 Gt C from fossil fuel consumption between 2000 and 2100. Cumulative carbon emissions for stabilisation profiles of 450 and 750 ppmv over the same period are between 630 and 1300 Gt C. Fossil fuel scarcity, at least at the global level, is therefore not a significant factor in considering climate change mitigation.

A major change of the energy supply system to replace fossil fuels gradually by other primary energy resources will be required. Combating a human-induced climate change becomes a central issue in the context of societal development and sustainability in general.

The IPCC therefore made substantial efforts to consider the issue of mitigation on the basis of the scenarios for the future development in the world, based on the IPCC scenario emissions report (IPCC, 2000b). It is, however, obvious that the reliability of the information that can be deduced on the basis of such scenarios declines rapidly after only a few decades and the economic aspects of alternative future energy supply systems were therefore still very unclear. Basically only the total range of future greenhouse gas emissions that the SRES describes can be relied on beyond the middle of the century.

The report describes the sensitivities of the scenarios to the assumptions made in their construction, i.e. population increase, economical development, technological innovations, etc. In addition, however, there will be major political consequences as countries try to secure their future supplies of energy during a transition period before other forms of primary energy, particularly renewable energy, become available in adequate amounts. Because of the huge differences in economic strength between industrialised and developing countries, this competition for resources will not favour equity and sustainable development in poor, developing countries. It is of course not the task of the IPCC to address the political difficulties that will most likely arise, but access to the best possible factual background is most desirable in this context. It is, however, not easy to bring together relevant information. For example, as I pointed out in my presentations to the Convention in 1997, the average per capita use of fossil fuel energy was about five times larger in industrialised countries than in developing countries during the last years of the twentieth century. What will this mean in the long-term perspective?

It is not possible to describe the content of the Working Group III report on a page or two, and I refer the reader to the summary for policy makers and the synthesis report for more detailed consideration. The extracts from the summary for policy makers that follow have been selected to be of particular interest to the development of strategies for mitigation to be agreed by the Climate Convention. Admittedly the choices are subjective, but may catch aspects of the report that will be of interest in the later discussion.

(7) Significant technical progress relevant to greenhouse gas emissions reduction has been made since SAR in 1995 and has been faster than anticipated.

(8) Forests, agricultural lands, and other terrestrial ecosystems offer significant carbon mitigation potential. Although not necessarily permanent, conservation and sequestration of carbon may allow time for other options to be further developed and implemented.

(10) Social learning and innovation, and changes in institutional structure could contribute to climate change mitigation.

(11) Estimates of costs and benefits of mitigation actions differ because of (i) how welfare is measured, (ii) the scope and methodology of the analysis, and (iii) the underlying assumptions built into the analysis. As a result, estimated costs and benefits may not reflect the actual costs and benefits of implementing mitigation actions.

(17) The successful implementation of greenhouse gas mitigation options need to overcome many technical, economic, political, social, behavioural, and/or institutional barriers which prevent the full exploitation of the technological, economic and social opportunities of these mitigation options.

(19) The effectiveness of climate change mitigation can be enhanced when climate policies are integrated with the non-climate objectives of national and sectorial policy development and be turned into broad transition strategies to achieve the long-term social and technological changes required both for sustainable development and climate change mitigation.

(20) Coordinated actions between countries and sectors may help to reduce mitigation costs, address competitiveness concerns, potential conflicts with international trade rules, and carbon leakage.

(22) This report confirms the findings in the SAR that earlier actions, including a portfolio of emission mitigation, technology development and reduction of scientific uncertainty, increase flexibility in moving towards stabilisation of atmospheric concentrations of greenhouse gases. The desired mix of options varies with time and place.

As might be clear from this brief summary, Working Group III struck a rather optimistic tone regarding the technical possibilities of making serious efforts to mitigate climate change and emphasised the modest costs in a long-term perspective for doing so, but considered also the number of barriers that have to be overcome, particularly the inertia due to social attitudes, the existing infrastructure, and not least conservative responses from existing institutions. However, a major and very difficult step is still required to transform these rather general conclusions into clear strategies for how to instigate future mitigation. At the time few were as yet engaged in such efforts.

11.2 Resistance towards taking action and political manœuvring

Major political difficulties arose soon after the agreement on the Kyoto Protocol in 1997 and these resulted in slow progress for almost a decade. Even though the EU took some significant steps towards realising the intentions as expressed in the Protocol, it soon seemed likely that the goals as formulated for the commitment period 2008–2012 would not be reached. I will not attempt to describe in detail the way the situation evolved, but rather limit myself to some key observations.

It is ironic that when in autumn 2000 the IPCC prepared for the final working group sessions to consider and approve the TAR, George W. Bush won the presidential election in the USA and on 13 March 2001 he wrote a letter to Senators Hagel, Helms, Craig and Roberts:

My administration takes the issue of global climate change very seriously . . . I oppose the Kyoto Protocol because it exempts 80 percent of the world, including major population centres such as China and India, from compliance, and would cause serious harm to the US economy . . . I support a comprehensive and balanced energy policy that takes into account the importance of improving air quality . . . I intend to work with Congress on a multiple strategy to require plants to reduce emissions of sulfur dioxide, nitrogen oxides and mercury . . . I do not believe, however, that the government should impose on power plants mandatory emissions reductions for carbon dioxide . . . A recent Department of Energy Report concluded that including caps on carbon dioxide emissions as part of a multiple emissions strategy would lead to even more dramatic shifts from coal to natural gas for electric power generation and significantly higher electricity prices . . . This is important new information that warrants re-evaluation especially at times of rising energy prices and serious energy shortage . . . we must be very careful not to take actions that could harm consumers. This is especially true, given the incomplete state of scientific knowledge of the causes of and solutions to global climate change . . .

The letter is a confusing statement and in no way a clear strategy to combat climate change. It rather expresses a wish to avoid criticism from industry and consumers and shows no sign of taking the global climate change seriously.

The US withdrawal from the Kyoto Protocol was a severe set-back and meant that the resumption of the sixth conference of the parties to the Climate

Convention scheduled for mid-2001 became rather meaningless. It was also obvious that the USA would have had considerable difficulties in fulfilling its obligations as specified in the Protocol. From 1990 until 2001 its carbon dioxide emissions increased by almost 15%, while the US obligation according to the Protocol was a reduction by 7% by 2008–2012. Few if any measures had so far been taken. The US Senate demand for developing countries to take on compulsory commitments hamstrung the US representatives to the conferences of the parties. Thus, the prospects for an early ratification of the Kyoto Protocol were slim.

In a letter of 11 May 2001 The White House asked the US NAS for assistance in identifying the areas in the science on climate change where there are greatest certainties and uncertainties. The NAS was also asked for its views on whether there are any substantive differences between the IPCC reports and the IPCC summaries. An answer to the request was expected in early June, i.e. within less than a month. The NAS quickly appointed a special committee under the chairmanship of Dr Ralph Cicerone, chancellor of the University of California, Irving, CA, and a well-known researcher in atmospheric chemistry (and president of the NAS since 2005).[11] Its report was ready in June and the following general conclusion was drawn (National Research Council, 2001):

The committee generally agrees with the assessment of the human-caused climate change presented in the IPCC Working Group I scientific report, but seeks to articulate more clearly the level of confidence that can be ascribed to these assessments and the caveats that need to be attached to them. This articulation may be helpful to policy makers as they consider a variety of options for mitigation and/or adaptation.

The committee finds that the full IPCC Working Group I report is an admirable summary of research activities in the climate science, and the full report is adequately summarised in the Technical Summary.

The committee itself then summarises briefly our present knowledge about key issues with the US perspective in mind, discusses the IPCC operational set up and procedures in order to answer the specific questions that were asked in the latter part of the letter. The following are some of the key responses by the committee:

The committee finds that the conclusions presented in the SPM and the Technical Summary are consistent with the body of the report. There are, however, differences. The primary differences reflect the manner in which uncertainties are communicated in SPM . . . This is perhaps understandable in terms of the process in which the SPM attempts to underline the major areas of concern associated with a human-induced climate change.

Human decisions will almost certainly alter emissions over the next century. Because we cannot predict either the course of human population, technology, or social transformations with any clarity, the actual greenhouse gas emissions could be either greater or less than the IPCC scenarios. Without an understanding of the source and degree of uncertainty, decision makers could fail to define the best ways to deal with the serious issue of global warming.

The SPM results from a discussion between the lead authors and government representatives (including also some non-governmental organisations and industry representatives). This discussion, combined with the requirement for consistency, results in some modifications of the text, all of which are carefully documented by the IPCC. This process has resulted in some concern that the scientific basis for the SPM might be altered . . . Based on an analysis, the committee finds that no changes were made without the consent of the convening lead authors and that most changes that did occur lacked significant impact. However, some scientists may find fault with some of the technical details, especially if they appear to underestimate uncertainty.

These comments of course refer to the controversies that followed after the publication of the IPCC SAR in 1996. Suspicions that members of the Working Group I bureau might have manipulated the text (see Section 9.2) were presumably eliminated by this evaluation. It is also noteworthy that the arguments from the US Administration for not accepting the commitments as prescribed in the Kyoto Protocol thereafter were on political grounds rather than because of any scientific inadequacy of the work by the IPCC.

There were strong reactions from a number of industrialised countries, particularly from within the EU, to this unilateral action by the USA to withdraw from the Kyoto agreement. The Swedish Prime Minister Göran Persson, chaired the EU Council during the first half of 2001 and on behalf of the EU he urged President Bush to reconsider the position taken by the USA and concluded his letter by saying:

I would . . . like to emphasize that to the EU an agreement at the resumed session of COP-6, on the basis of the Kyoto Protocol that would lead to real reductions in greenhouse gas emissions, is of utmost importance. The global and long-term importance of climate change and the need for a joint effort by all industrial countries in this field makes it an integral and important part of relations between the USA and the EU. A resumed COP-6 is therefore urgently needed and I would like to express the wish of the EU to initiate such a dialogue at the highest level as soon as possible.

As is well known this appeal was in vain. The frustration felt by many did, however, have some effect and in February 2002 President Bush announced a climate change initiative that implied a reduction in the greenhouse gas intensity of US industry of 10% during the next ten years, i.e. industry would become more efficient in its use of energy. The Netherlands Environmental Assessment Agency, however, quickly showed that this rather ad-hoc approach would probably not lead to a significant reduction of the emissions (see de Moor *et al.* (2002)). I quote the following conclusions:

The policy target of 18% improvement in the greenhouse gas intensity of the US economy 2002–2012 is comparable to historical trends. In the 1980s the intensity in the US declined by 23% and in the 1990s by 17%. The intensity improvements . . . over the next 10 years in the Bush initiative are 14% . . .

The policy target in the Bush climate change initiative will result in a continued increase of US greenhouse gas emissions. In 2012 the emissions will be 32% above the 1990 level. This is far above the original Kyoto target for the US of -7% in 2010.

It is obvious that President Bush's proposal was merely to continue 'business-as-usual', but this was hidden in a wording that gave the impression of a move towards meeting some of the Kyoto goals, while its real implications were strikingly negative. Other campaigns, one under the title, 'Redefining Progress', also got under way as reactions to the initiative by President Bush.[12] The following call was widely distributed:

Recent Bush Administration announcements, including President Bush's press conference statement today that curbing carbon dioxide emissions would harm our economy and hurt our American workers ignore a declaration endorsed in 1997 by 2500 economists, including eight Nobel Laureates, stating that policies to slow climate change can be enacted without harming either the United States economy or living standards. As economists, we believe that global climate change carries with it significant environmental, economic, social, and geopolitical risks, and that preventive steps are justified. Economic studies have found that there are many potential policies to reduce greenhouse-gas emissions for which the total benefits outweigh the total costs. For the United States in particular, sound economic analysis shows that there are policy options that would slow climate change without harming American living standards, and these measures may in fact improve US productivity in the longer run.

But the supporting analyses were missing and this initiative, even if supported by 'eight Nobel Laureates' did not have much of an impact. It became newspaper headlines for a few days, but was then largely forgotten as quickly as it had appeared. A transparent process based on the published scientific literature and direct confrontations between politicians and scientists is the approach that in the long run will secure attention for the key issues and progress. After all, climate change will be with us for decades to come and continuity in the assessment process is a necessity. Nevertheless, it is remarkable that the only superpower and simultaneously the most prosperous country in the world and responsible for about 40% of the emissions from industrial countries, did not accept a task that most other industrial countries were willing to do.

11.3 Other challenges of the IPCC conclusions

In parallel with the political process there were efforts to discredit the IPCC's assessments of the available scientific knowledge. Some scientists expressed disbelief in the IPCC conclusions and thereby indirectly supported the lobbyists who feared the stringent measures that might be required in order to mitigate climate change, which some believed was an as yet unproven hypothesis. Such lobbyists were successful in delaying efforts to prevent climate change during the early years of the twenty-first century and thereby also hampered work towards sustainable development.

Other scientists associated themselves with environmental organisations because of their anxiety about the consequences of a global climate change. With such scientists it was not always possible to distinguish between scientifically credible conclusions and advocacy for urgent action as response to the serious threats that they felt global climate changes implied. The public debate became increasingly polarised.

Among the critics that intervened a few had excellent academic records and their arguments deserve serious consideration. One of the more vocal ones that attracted considerable attention was Richard Lindzen, professor at MIT in Cambridge, MA, USA. He focused his research efforts on the detection of negative feedback mechanisms, such as those due to the role of increasing concentrations of water vapour and clouds in a warmer world, rather than positive ones. This is in itself a legitimate scientific approach. He discovered an interrelation between intensified vertical convection due to the warming of ocean surface waters and the extension of cirrus clouds in tropical regions (Lindzen, 1997; Lindzen *et al.*, 2001). Since high level clouds reduce the incoming solar radiation, an increase in their amounts might counteract the enhanced warming due to increasing concentrations of water vapour in the atmosphere. Although the physics of the argument in principle is plausible, data to support the idea were obtained from a limited area in the eastern parts of the Pacific Ocean around Indonesia, where this kind of relationship was most likely to be found. The relationship postulated was only weakly supported by data from this area and nobody else has been able to ascertain that the process is in reality of significance. Nevertheless, the role of clouds in the ongoing changes of the global climate is poorly understood and deserves further attention.

It is interesting to note that proponents of such dissenting views often came from other fields of science, particularly the geosciences and solar physics. For example, an article by a well-known Canadian geologist (Veizer, 2005) illustrates the differences of methodological approach between different scientific fields. Veizer had analysed climatic data that span more than four billion years of earth's history. He based his conclusions on graphical representations of observed changes, correlations and analogues between different epochs, and largely disregarded the differences between slow climate changes over millions of years and the present very rapid changes occurring within a century or two. He also dismissed the elaborate theoretical analyses of present changes deduced with the aid of complex physical models, which of course can seldom be applied in paleo-climatology because of the sparse data sets that are available for validation in comparison with the massive amounts of data describing recent changes: see for example Simmons *et al.* (2004). His analysis is therefore inadequate.

Svensmark has found a correlation between variations of radioactive isotope Be-7 in the atmosphere (generated by cosmic rays) and cloudiness (see Svensmark *et al.* (1998) and Marsh and Svensmark (2003)). No plausible physical explanation has been put forward, but the observations were taken as a sign of a significant impact of cosmic rays on the climate system. However, there are almost always adequate numbers of particles in the atmosphere for condensation to occur as the humidity approaches saturation, since the relative humidity never significantly exceeds 100%. Why should cosmic rays then play a role? The correlation between clouds and the atmospheric concentration of Be-7 as found by Svensmark might primarily depend on the increased rate of wash out of the aerosols (to which the Be-7 atoms are attached) in cloudy and sometimes rainy weather compared to when skies are clear. The decisive correlation should therefore be between the amount of cloud and the neutron counts that register the formation of Be-7 atoms. Harrison and Stephenson (2006) have found that this correlation is rather weak and hardly significant.

These are a few examples of analyses that attempt to widen the search for mechanisms other than the enhanced greenhouse effect that may explain the ongoing global warming, though these have not as yet been validated. Whilst a readiness to accept divergent views must always be maintained, it is always necessary to examine such views critically.

A more elaborate, but still scientifically inadequate, effort to express doubts about the IPCC efforts was made by the Danish statistician Bjorn Lomborg, who published the book *The Sceptical Environmentalist* in 2001, of which one chapter dealt with the climate change issue. I will consider this work further because it so clearly shows how a detailed presentation of a scientifically and technically complex issue, can temporarily have considerable influence on the political debate even though it may be selective and biased and does not provide the scientifically thorough analysis that it pretends.[13] The introductory part of the chapter on climate change provides a fair overview of the basic features of the climate system. The more detailed presentation of the global warming issues is, however, biased and often plainly wrong. *Scientific American* referred the book to four well-known US scientists in various fields of the relevant sciences for careful review (see Schneider *et al.* (2002)). I largely share the gist of their analyses.

Stephan Schneider, a well-known climatologist and senior fellow at the Institute for International Studies at Stanford University, emphasised the following major shortcomings:

Lomborg considers, with no acceptable arguments, only the IPCC scenario for future emissions that result in the smallest IPCC emission scenario, i.e. scenario B1.

He misunderstands the use of a set of alternative scenarios aimed at exploring the sensitivity of the climate system to external forcing.

Although the climate change projections are uncertain, Lomborg maintains with inadequate arguments that the sensitivity of the climate to changes of greenhouse gas concentrations and aerosols in the atmosphere is at the low end of the IPCC uncertainty range. He simply states that the 'temperature will increase much less than the maximum estimates from IPCC – it is likely that the temperature will be below or at the B1 estimate' (i.e. well below $2\,^{\circ}$C in 2100).

In his cost-benefit analysis Lomborg does not consider the inadequacy of the assessment of the benefits of avoided climate change, and he provides no uncertainty estimate for the probably much too low value that he accepts as being most appropriate.

Lomborg misinterprets the Kyoto Protocol by ignoring that the proposed emission reductions only concern the time period until 2012.

All these issues were dealt with at length in the IPCC reports and by the Climate Convention when the Kyoto Protocol was agreed. One wonders how well Lomborg had read and appreciated those analyses. It is characteristic that most of his 600 references are found in the secondary literature and media articles. Schneider summarises his review by the sentence: 'Lomborg admits, "I am not myself an expert as regards environmental problems" – truer words are not found in the rest of the book.'

John Holdren, professor of environmental policy at the John Kennedy School of Government, at Harvard University, gives a number of examples of Lomborg's misconception of the global energy situation, potential geopolitical controversies and his superficial view of available energy reserves and resources. He says that Lomborg ignores the difficult management problems that are likely to arise because of the major transitions that the world will have to go through in our striving for a future sustainable energy supply system. This has actually already begun. Some of Lomborg's own projections were already outdated five years after he completed his book. Socio-political issues which will be so decisive for the future were mostly ignored.

Efforts have been and are being made to slow down the increase in the population of the world (the implications of which were explored by the IPCC) and it is wrong to ignore this issue as Lomborg explicitly does. *John Bongaarts*, vice president of the Policy Research Division of the Population Council in New York City, did not mean that the population issue is the main cause of the world's social, economic and environmental problems, but . . . if population had grown less rapidly in the past we would have been better off now. And if future growth can be slowed, future generations will be better off.

Lomborg's way of ignoring central issues, about which considerable knowledge is available, is highly questionable in the context of outlining a sound future approach towards finding strategies for sustainable development.

Finally, *Thomas Lovejoy*, chief biodiversity advisor to the president of the World Bank and senior advisor to the president of the United Nations Foundation, characterises Lomborg's analysis of the issue of biodiversity as '. . . dismissing the scientific process.' He provides numerous examples of a selective choice of examples and numbers, and a set of very biased conclusions based on Lomborg's subjective interpretation of the state of knowledge of the phenomena being analysed.

Based on these four reviews of parts of Lomborg's book and my own reading I conclude that the book does not reach the required scientific standards. It misleads anyone not very familiar with the various issues. Scientific stringency is often lacking, and difficult to pinpoint without going back to the original publications to which he refers. Lomborg tries to get to grips with the very complex global issues that will confront mankind during the twenty-first century, but it is next to impossible for a single individual to write an overview of the kind that he has attempted without having access to overviews of most of the subfields by well-established scientists and a willingness to accept their conclusions. The scientific literature is simply too large for a single individual to grasp primarily by considering individual scientific papers. This general criticism is valid for many of the publications concerning global warming.

11.4 The leadership of the IPCC is changed

Robert Watson's five-year term as chairman of the IPCC expired after the completion of the TAR. The election of a chairman to lead the IPCC in its work towards a fourth assessment was put on the agenda for the nineteenth IPCC plenary session, held in Geneva in April 2002. A large number of scientists from the industrialised countries were in favour of Watson being reelected, not least those from the USA and he was most anxious to continue. He had led the IPCC work in a forceful and innovative manner. His broad knowledge about the related sciences had been instrumental in fulfilling the key task of an IPCC chairman, i.e. to maintain the independent status of the IPCC. I was in favour of his reelection. I was therefore surprised, when the USA did not nominate him, but rather proposed Dr R. Pachauri, a well-known researcher in the field of energy and head of TERI, in New Delhi, India. The move was obviously a political one, but this statement does not reflect negatively on Pachauri as a scientist.

The oil-producing countries had constantly been challenging Watson as they had often opposed me during my time as IPCC chairman. The US Administration generally gave priority to energy supply issues and also did so in the context of the climate change. Its relations with the oil-producing countries in the Middle East were also of prime concern. Watson, having been elected during the presidency of Bill Clinton and personally very engaged in the climate change issue, had presumably been found too independent by the new US Administration. The proposal that Dr Pachauri should take over was a clever move. He was well known and in many regards was a most competent representative from a developing country, while Watson's quick and straightforward approach to people sometimes made the representatives of these countries uneasy. However, Watson was nominated by the Portuguese and British delegations (in itself an unusual move) and Sweden was asked if I, as a previous IPCC chairman, might be willing to join the Swedish delegation and then serve as chairman for the nominating committee. I agreed to do so and took on this function at the Geneva meeting.

The discussions led to increased tension between developed and developing countries, which was taken advantage of by oil-producing countries and the Russian Federation. In the final vote Pachauri received 76 votes, Watson 49, and Jose Goldenberg (Brazil) 7. Pachauri's victory was primarily due to the overwhelming support he received from the developing countries who were anxious to secure more influence for themselves. It was also felt by some that Watson's high tempo and his management style left them behind. I knew, however, that he had a genuine wish to accommodate the interests of the developing countries. Also, the job of chairman of the IPCC is very demanding and I was concerned that Pachauri, who was also the executive director of TERI in India, might be overextended. I was disappointed about the outcome of the vote and felt that the intensity of the IPCC activities might decline. When leaving Geneva at the end of the meeting I accidentally met Bob Watson at the airport. He was cheerful as always but I felt strongly that he was greatly disappointed by the outcome of the election.

In any event, a change of leadership had been decided on and it was vital that this did not cause any lasting ill feelings amongst the scientists engaged in the IPCC work. In a note in *Nature* I expressed the following views (Bolin, 2002).

There is a need for a genuine spirit of cooperation between developed and developing countries to combat climate change . . . The required reductions of future global emissions of greenhouse gases will only be possible if the lead is taken by developed countries, as is clearly expressed in the Convention on Climate Change. This implies necessarily that the present very large differences in per capita emissions of carbon dioxide between countries must be reduced to secure sustainable development in developing countries and simultaneously to strive for reduction of global emissions (see Bolin and Kheshgi, 2001).

There were also other changes in the leadership of the IPCC on this occasion. Drs Dahe Qin, China, and Susan Solomon, USA, were elected cochairmen of Working Group I, Drs Osvaldo Canziani, Argentina, and Martin Parry, UK, cochairmen of Working Group II, and Drs Ogunlade Davidson, Sierra Leone, and Bert Metz, The Netherlands cochairmen of Working Group III. Susan Solomon was the first woman in the leadership of the IPCC and she had earned her position through her eminent work as an atmosphere chemist, especially in investigating the changes of the ozone layer. The work on a fourth assessment began. It was agreed that 2007 would be the target date for its completion.

11.5 Ratifications of the Kyoto Protocol

In spite of the more confident conclusions in the TAR about an ongoing global climate change, the Kyoto process came almost to a standstill. The Russian Federation, which in 1990 was responsible for about 17% of the industrialised countries' emissions also hesitated and delayed ratifying the Kyoto Protocol. The world oil and gas market was becoming increasingly important for the Russian Federation and the implications of ratifying the Protocol needed 'serious considerations', in the words of President Vladimir Putin. His refusal for a long time to ratify meant that USA, Australia and the Russia Federation effectively prevented the Protocol coming into force.

With support from the Russian Academy of Sciences and President Putin, Academician Yuri Izrael, vice-chairman of the IPCC, invited the countries of the world to a third World Climate Conference (WCC) in Moscow. It was announced as a follow-up of the second conference of this kind organised by the WMO in 1990, on which occasion strong support was given to the conclusions of the IPCC FAR, which had then just been completed. At the meeting in Moscow, in 2003, the aim of the leaders of the Russian Academy, Yuri Izrael and Kiril Kondratiev,[14] turned out to be the opposite, i.e. not to accept the conclusions drawn in the third assessment.

The conference took place between 29 September and 3 October 2003. I was asked to give the opening lecture at the Conference and chose to outline what we know quite well about the ongoing climate change and what we are therefore able to foresee with reasonable certainty, rather than primarily emphasise the uncertainties.[15] I stressed in particular the necessity of dealing with the north–south dimension of the issue. The economic advisor to President Putin, Professor Illiaronov, of Moscow University, was also asked to address the conference. He expressed his strong doubts about the IPCC reports and indicated very clearly that he did not believe that the Kyoto Protocol would achieve its aims. He asked for answers by the scientists to ten questions that he spelled out in his talk.

I had earlier been asked by Dr Izrael to chair the committee that was supposed to formulate a statement about the outcome of the conference for consideration at the closing session. I seized on this opportunity to respond to Illiaronov's questions, called on the committee to meet and asked Izrael to invite Illiaronov back to receive our response.

Eight of his questions were simply surprising misinterpretations of the text and graphs in the last IPCC report that could easily be rectified. The two last questions concerned whether or not the Kyoto goal for 2008–2012 was likely to be achieved and if the IPCC assessment of the cost estimates for concerted action in order to stabilize the carbon dioxide concentration in the atmosphere at the level of about 550 ppmv was reasonable. The likelihood of achieving the 5% reduction in greenhouse gas emissions by developed countries largely depended on whether or not the reductions of the Russian emissions since 1990 (the 'hot air') would be enough to compensate for the increases of emissions by other industrial countries, particularly the USA. The response to the very last question was basically straight forward: in comparison with the expected future increase in the gross domestic product of developed countries the costs would be modest, even small, although the total costs might add up to trillions of dollars (10^{12}) for the twenty-first century as a whole. The annual increase of the gross domestic product has been on average about 2% per year for most countries during the last few decades. Merely setting aside a small part of this increase would furnish financial resources that might suffice to mitigate the ongoing changes with the aim of keeping the carbon dioxide concentration from doubling.[16] Dr Illiaronov did not, however, accept the simple and straightforward analyses with which he was presented. From his perspective issues of economic development in the near future were obviously much more important than long-term environmental issues. When the conference closed later in the day, he gave a press conference in another part of Moscow, again criticising the IPCC reports and the Kyoto Protocol. He was obviously supporting the Russian delay of the ratification of the Protocol, but it was not clear what influence he might have on decisions by President Putin.

Dr Izrael also organised an ad-hoc open debate, to which he invited only sceptics of a human-induced climate change as speakers, among them Professor Lindzen, USA, and Dr Sonja Böhmer-Christiansen, UK, who took part in the discussion through video telephone arrangements. Izrael criticised the IPCC conclusions without reference to any specific analyses in the scientific literature and as moderator of the session did not permit the audience to become engaged in a serious exchange of scientific arguments. The setup was strange in that

Izrael did not argue on the basis of the conclusions of the IPCC reports. He put forward his own personal views, which were a mix of scientific and political arguments that confused the audience rather than clarified the essence of the IPCC reports, to which he still referred. The debate was announced as a scientific one but in reality became a political crusade by Izrael. The audience was quite upset, a number of Russian journalists protested strongly and the debate ended in chaos.

It should also be added that the Russian Academy sent a formal declaration to President Putin in the spring of 2004 advising against Russia ratifying the Kyoto Protocol.[17] The logic of the document is, however, difficult to follow, not least because of an inappropriate reference to my presentation at the Moscow conference. Several of the statements on future emissions are simply wrong. For example, the declaration claimed that it was very uncertain whether global warming is produced by human emissions and that the natural carbon cycle is not well understood because of a lack of data. The analysis of possible impacts of a climate change is brief and unsatisfactory. It does not recognise that the Kyoto Protocol is merely the first step towards the stabilisation of the atmospheric carbon dioxide concentration. Rather, the declaration says that the Kyoto Protocol would be ineffective in achieving the aims of the UN FCCC. The document further emphasises that it would be difficult to achieve President Putin's vision of doubling Russia's gross domestic product during the next ten years (i.e. an annual increase by about 7%) if Russia were to accept and ratify the Kyoto Protocol. Nowhere is it mentioned that the Kyoto Protocol allows Russia to increase its emissions until 2008–2012 back to their level in 1990, an increase of about 30%, which none of the OECD countries enjoys. In addition, the document complains that the planned mechanisms of the Kyoto Protocol give advantages to developing countries at Russia's expense and that the temperature regime in Russia, the coldest country in the world, was not taken into account.

One wonders how these could possibly be accepted as legitimate arguments in the political process within the Climate Convention. This declaration by the Russian Academy is indeed a most amazing document, a mixture of poor scientific arguments and national politics. It is surprising that this kind of document was presented by the leading scientific institution of the country.

Later in the year representatives of the British Government invited their Russian counterparts, still under the leadership of Izrael, to discuss the topic of how to deal with the issue of mitigating climate change. However, without any prior announcement, Izrael brought Professor Lindzen with him to the meeting. The British delegation failed in their hope of convincing the Russians of the seriousness of the climate issue.[18] In spite of the almost unanimous agreement amongst leading scientists in the world as summarised by the IPCC, both the US and the Russian governments were still not ready to accept its conclusions. The reasons were obviously purely political.

In any event, the EU proceeded with its preparations for the legal acceptance of the Kyoto Protocol. The charge of reducing emissions by 8% had been modified with regard to the distribution of responsibilities between member countries in that some accepted more stringent requirements, while others were given less demanding commitments. A proposal for the distribution, 'burden-sharing', was prepared by the Dutch national committee based on an assessment of the particular circumstances of each country of the EU and was unanimously agreed. The EU also tried hard to convince other countries to ratify the Kyoto Protocol in order to ensure its coming into force as soon as possible.

In spite of the advice given by the Russian Academy of Sciences, the Russian Federation finally announced its ratification of the Kyoto Protocol in November 2004 and the treaty came into force in February 2005, almost ten years after the first conference of the parties in Berlin.

11.6 The eleventh conference of the parties to the Climate Convention

Toronto, Canada, was the venue of the eleventh conference of the parties to the Climate Convention in December 2005. The session was dominated by the fact that the Kyoto Protocol had come into force. One after another, the heads of delegations to the session spoke in glowing terms about the progress that had been made, while the USA maintained a cautious and rather low profile. Representatives of environmental groups expressed their optimism about the future, but also their unease about the fact that progress still was slow. In order to emphasise further the importance of the occasion the Canadian delegation extended an impromptu invitation to the former US President Bill Clinton to address the gathering. He accepted the call in spite of the short notice, but was careful to stay away from the political controversies that obviously lurked beneath the spoken words at the conference. He instead emphasised strongly the need for further technical development in order to deal with the many difficult issues that remain to be resolved and rightly so.

The disappointing feature of the conference was its inability to consider more specifically the most important issue now to be addressed by the convention: what commitments should be agreed for the commitment period that will begin in 2013? A task force was appointed to negotiate and put forward a proposal to the thirteenth session of the convention that will take place in 2007. Obviously time is short and controversies will certainly emerge because the USA is not a party of the Protocol, and nor is Australia.

It is surprising and sad still to find considerable resistance in the US Congress to the reality of the global warming. Senator James M. Inhofe, chairman of the Committee on Environment and Public Works of the US Senate, in November 2005 returned to the discussion about the IPCC approval of the SAR in 1995.[19] He aired again the misconceptions that I have dealt with in detail in Chapters 8 and 9 and he completely disregarded the report by the US NAS in June 2001 to President George W. Bush, which expressed its high regard for the work by the IPCC.

Even worse, he disregarded (or was possibly not aware of) the statement made by the presidents of the eleven most knowledgeable academies of science in the world, i.e. those of the G8 countries and three major developing countries: Canada, France, Germany, Italy, Japan, Russia, United Kingdom, USA, and Brazil, China and India. Their conclusions for action were very much along the lines I have pursued in my analysis here and deserve to be repeated:

- Launch an international study (recognising and building on the IPCC's ongoing work on emission scenarios) to explore scientifically-informed targets for atmospheric greenhouse concentrations and their associated emissions scenarios, that will enable nations to avoid impacts deemed unacceptable.
- Identify cost-effective steps that can be taken now to contribute to substantial and long-term reduction in net global greenhouse gas emissions. Recognise that delayed action will increase the risk of adverse environmental effects and will likely occur at greater costs.
- Work with developing nations to build a scientific and technological capacity best suited to their circumstances, enabling them to develop innovative solutions to mitigate and adapt to the adverse effects of climate change, while explicitly recognising their legitimate development rights.
- Show leadership in developing and employing clean energy technologies and approaches to energy efficiency, and share this knowledge with all other nations.
- Mobilise the science and technology community to enhance research and development efforts, which can better inform climate change decisions.

This declaration was a welcome support of the IPCC's efforts to bring the climate change issue to the forefront of the political debate. Most appropriately, the IPCC has just completed its Assessment Report 4 (AR4) and an up-to-date description of present knowledge has become available (IPCC, 2007a,b,c). Some uncertainties in the third assessment have been reduced, but there are fundamental uncertainties that will remain and that cannot be eliminated. The predictability of the future is limited and the fundamental question will undoubtedly be: how can we find pathways towards the future that avoid the major pitfalls and that are still socially and economically acceptable, if not optimal, for the countries of the world?

However, in the broad political context the following questions remain. How urgent is it to act now rather than later? Is it possible to find a way to progress the negotiations further? How might science best contribute to stimulating progress in this regard? In any case, future changes of climate are certainly going to be with us for many decades to come, probably for a century or even more. Adaptation measures are therefore needed, but the main concern will be mitigation in order to keep the ongoing changes within bounds. Some initiatives are being taken, but the pace is still slow. Nevertheless, the year 2006 may indeed have been a turning point with regard to the human-induced climate change issue. The IPCC's AR4 is being published and summarises the scientific progress during the last five years. Chapters 12 and 13 give my views on the opportunities and constraints that humankind is facing.

Part Three

Are we at a turning point in addressing
climate change?

12

Key scientific findings of prime
political relevance

The reality of a human-induced climate change is becoming more generally accepted. Preparations for adaptation have begun. The Kyoto Protocol has come into force, but no long-term agreement on mitigation has yet been reached.

12.1 The general setting

The early eagerness amongst politicians around 1990 to act in response to the threat of a human-induced climate change was largely genuine and in line with the increasing general attention that was given to environmental issues during much of the 1990s. There was, however, early reluctance from industry and other stakeholders to proceed quickly. They feared that action to protect the current climate, i.e. a reduction in the use of fossil fuels, might be a threat to their activities and admittedly the scientific basis for taking action was then hardly convincing.

Other major global issues, particularly many of a political nature, have since been brought into focus and have greatly influenced world politics, especially since the turn of the century. The conflicts in the Middle East to some considerable degree stem from a realisation that the global energy supply system might have to change during the coming decades. The conventional reserves of oil will dwindle within a decade or two and natural gas will begin to run out towards the middle of the century, while the global demand for energy will be increasing quickly, not least because of the rapid industrialisation in developing countries.

There is a need for trustworthy scientific information in order to find a common strategy between, on one hand, those that are giving priority to the *short-term* political security in today's society and, on the other, those that are anxious to safeguard the global environment and obtain a sustainable development emphasising the *long-term* issues. Of course, both aims deserve attention.

Similar polarisations of conflicting views have in the past developed about other environmental issues. The fish stock in the oceans has decreased rapidly over several decades because of over-fishing, in spite of the scientific efforts to clarify the critical long-term development that is on its way. The situation is still deteriorating. Scientific advice has not been taken seriously. Should we fear a similar situation with climate change?

On the other hand, the threat to the ozone layer has been dealt with successfully, so far at least, though here the measures required were much less costly and the problem was perceived by people in general as a much more imminent threat. The words 'cancer due to enhanced UV-radiation from the sun because of a hole in the ozone layer' are a powerful argument in discussions with the public. The threat to life was felt to be immediate, rather than in some distant future and 'elsewhere' on the globe. Few were disturbed by the measures that were required and these have largely been implemented.

The need for an increasing supply of energy and the long-term aim of improving conditions for life in developing countries have long posed acute problems for humankind and must be given priority. This must necessarily be coupled to the threat of a climate change if fossil fuels were used as the future prime source of energy for the world. Is it then possible to ensure that the demands for more energy services in the world of tomorrow can be met without an excessive use of fossil fuels and a serious risk of global climate change?

12.2 The story of global warming told to politicians, stakeholders and the public

12.2.1 To deal with controversies

It should be clear from the presentations in the previous chapters that accepting the IPCC conclusions about ongoing changes of the global climate as the starting point for developing a strategy for the future is well justified. The appearance of the summary for policy makers of the AR4 in early 2007 (see IPCC (2007a,b,c)) further supports this view, and also brings with it interesting new results from ongoing research efforts, in particular because of a much richer set of experiments with global climate models. Nevertheless, we all know that projections into the future cannot be checked against observations and some basic processes and secondary feedback may still be poorly described. An obvious deficiency is, for example, feedback from the terrestrial system as it changes because of the human-induced climate change is not included, and there may well be other internal interactions that should be dealt with better. Projections beyond the middle of the century can only be relied upon with regard to their large-scale features. However, the AR4 gives more information about uncertainties in projections of the future by comparisons with results from several models, which call for probability analyses.

Another development in early 2007 also deserves comment. A group of climate researchers, who on earlier occasions expressed their doubts about the IPCC procedure and its conclusions, have formed a writing team to present an 'Independent Summary for Policymakers – IPCC Fourth Assessment Report'.[1] The ten researchers have primarily been engaged in data analysis and refer to a number of publications in renowned scientific journals. It is an attempt to bring together their findings in the perspective of the IPCC effort, but deals only partly with the report from IPCC Working Group I and primarily analyses of past changes of the climate system. It has been carefully done but is still biased in the choice of results it emphasises. The draft report has been reviewed by other researchers, many of whom have been sceptical about the IPCC achievements. The review process was hardly unbiased. Still, it might be used as a starting point for less controversial discussions than in the past between rather widely separated groups of researchers.

The report rejects the IPCC view that 'It is *very likely* that anthropogenic greenhouse gases increase has caused most of the observed increase in globally averaged temperatures since the mid-twentieth century' (see IPCC (2007a)). The scientific justification for this rejection is not convincing in the light of the detailed analyses, in particular on the basis of a large number of modelling experiments, that have appeared in the scientific literature since the turn of the century. The statement in their report that 'computer simulations . . . can never be decisive as supporting evidence' is of course formally correct in that it cannot be strictly proven that they tell the exact course of future events, but the IPCC conclusions are still very plausible, being based on the wide variety of model experiments that have been completed. The very last sentence in the group's executive summary is of course correct: '. . . there will remain an unavoidable element of uncertainty as to the extent that humans are contributing to future climate change, and indeed whether or not such change is a good or bad thing . . .' provided that the last few words are changed to '. . . indeed to what extent such changes would be bad or exceptionally even might be beneficial locally.' The key conclusions in AR4 largely remain unchallenged.

Policy makers would also have major difficulties in interpreting the analysis by the group in terms of what to do. The outcome is in reality an implicit recommendation to do nothing until better scientific advice can be provided. The implications of such an attitude are far-reaching because of the inertia of the climate system and of the global society, which ultimately will have to deal with a future change, regardless of it being modest or large. The interplay between induced changes of the natural system and the global socio-economic system is not at all addressed in the group's report.

It should be clear from the presentations in previous chapters, further reinforced by the IPCC AR4, that focusing on the development of a strategy for the future based on the IPCC conclusions is now well justified. But in spite of the rapid increase in the general awareness of the climate change issue, it is legitimate to ask the questions: How well are key facts about climate change actually understood by politicians and the public in general? What is particularly important to know and to appreciate in order to accept the need for the development of a long-term climate policy now? How can better awareness of the most urgent issues be spread widely? How can we ensure that the more penetrating analyses of available scientific knowledge are taken seriously? How should they be presented to be helpful in a political context?

There is still a gap in knowledge between the scientific community directly engaged in the climate change issue on one hand and journalists, politicians, the general public and even a few other scientists on the other. How many of the latter groups have even read the IPCC summaries for policy makers carefully, undoubtedly the most relevant and also most easily accessible output from the IPCC? I am afraid that this question has a disappointing answer. Even colleagues in other fields of science are often poorly informed and dry scientific prose is not what people in general turn to in order to be kept informed about issues outside their own profession. Rather, stories told in newspapers, television or films attract people's attention. One journalist expressed this by saying that journalists prefer to read what other journalists write!

Basic scientific knowledge rather must be told to the general public as a story that brings the threats to human life and well-being into the forefront, not merely as an enumeration of numbers and graphs that describe findings and projections about the future. But scientific integrity must not be compromised. The following paragraphs then need particular attention.

Robust scientific findings must be underscored in a manner that recognises the urgency to resolve key political issues. Different views should be challenged scientifically to prevent the spreading of misinformation and unnecessary delays in taking appropriate measures.

Part of the problem is that the details of the regional and, in particular, local features of human-induced climate change cannot be described in very specific terms. Only the gross features are reasonably reliable. Expected regional and, especially local changes must rather be provided in terms of risk panorama. This is not an easy task and has seldom been done until quite recently.

Although there may also be winners as a result of climate change, no country will be unaffected by the dominating negative impacts. Our natural surroundings as well as society have been tuned to the climate that has prevailed for centuries. Adjustment to a different environmental setting is not simple. It might cause social unrest and be costly if it has to be achieved quickly. More rapid policy changes and more costly interventions might later become necessary if the present slow progress of the international negotiations continues.

There is an obvious need to develop new energy supply systems for the world, not only in order to slow down and ultimately stop the increase in atmospheric carbon dioxide concentrations, but also because of the expected reduced availability of cheap fossil fuels during the twenty-first century. What does this really imply? To be successful in such an effort it is necessary that cooperation between all countries of the world is fostered, and similarly cooperation between scientists, politicians, stakeholders in industry, agriculture and forestry, as well as the public at large.

Lasting political solutions cannot be found unless a development towards a more equitable world, based on democratic values and institutions takes place. This process has begun but progress is slow. To slow down and ultimately stop a global climate change is a prerequisite for long-term sustainable development.

An effective interplay between scientists and government representatives engaged in the climate change issue has been achieved during the last 15 years through the IPCC efforts. Its analyses have been decisive in providing an authoritative presentation of available knowledge and have played a crucial role in the establishment of a scientific basis for the Climate Convention and the Kyoto Protocol. I feel, however, that a key stumbling block today is the fact that the scientific community does not yet fully appreciate the way politicians make use of and the general public interprets the information that scientists provide. The full story of climate change must be told in a simpler but still trustworthy way than has been the case in the past. Exaggerated media descriptions of the threats do not pave the way for constructive cooperation. The following analysis is an attempt to pursue this kind of approach and is focused on measures that need to be taken, but my effort certainly needs further elaboration and should not replace detailed scientific analyses. Thus the following is *my* attempt to bring together the knowledge that I have gained and the experience that I have had in my contacts with politicians, stakeholders and the public in general. It finally boils down to an *evaluation of the relative importance of different features of available knowledge that necessarily also implies value judgements by the recipient of the information.*

12.2.2 Global scale climate changes

The following IPCC conclusions concerning the climate system need to be considered in political negotiations. They take into consideration the IPCC conclusions from 2001 and the summary for policy makers of the IPCC's AR4.[2] *The points made are all quite robust and should not be scientifically controversial.*

On the time scale of a human individual the climate system is responding to human interference rather slowly, but the response is fast seen from a geological perspective. Gradual changes due to natural geological processes can therefore largely be ignored when considering the human-induced changes of the climate system that are now being observed. There are, however, still some disagreements about the relative importance of natural variability and human-induced changes on the regional and local scale that need to be resolved.

The inertia of global society is considerable. Its infrastructure has been built over a period of one to several centuries. Seen in this context the response of the natural systems to the present imposed disturbances is rather rapid and persistent, while the implementation of measures to adapt and mitigate might take half a century or more. At present they have barely begun.

The global mean surface temperature has so far (by 2007) risen by $0.8 \pm 0.2\,°C$ since pre-industrial times (see Figure 12.1(a)).[3] It has risen two-tenths of a degree since the TAR was published six years ago. Over the continents, where humans live and work, the average global warming is on average about $1.0\,°C$, over the oceans it is as yet only a little more than half a degree, but regional variations might be $\pm 1\frac{1}{2}\,°C$. The Northern Hemisphere has warmed more than the Southern one. The global mean temperature has not increased regularly during this time. It rose by about $0.3\,°C$ until the mid-1930s, probably primarily due to a temporary small increase in solar radiation and the almost complete absence of volcanic eruptions from about 1910 to the early 1960s, while the increasing greenhouse gas concentrations would only have played a rather marginal role before about 1930.

The rapidly increasing use of fossil fuels as a prime source for energy in the process of industrialisation led to a more rapid increase in greenhouse gas concentrations in the atmosphere after World War II, although air pollution (particularly emissions of sulphur dioxide) also increased and counteracted the warming.

The carbon dioxide concentration in the atmosphere has now increased by 36% since the early nineteenth century, i.e. from about 280 ppmv to about 305 ppmv in 1940 and above 380 ppmv today, an increase that exceeds anything that has been observed during the last about 650 000 years. It is virtually certain that this is primarily the result of burning fossil fuels, but deforestation and changing land use have also contributed significantly. In addition the concentrations of other human-induced greenhouse gases have also increased and the total enhanced greenhouse gas forcing is equivalent to an increase in carbon dioxide to about 450 ppmv, The contributions to the enhanced greenhouse effect by all radiatively active components in the atmosphere as of 2000 are summarised in Figure 11.2. The increase in greenhouse gas concentrations during the last six years implies an

CHANGES IN TEMPERATURE, SEA LEVEL AND NORTHERN HEMISPHERE SNOW COVER

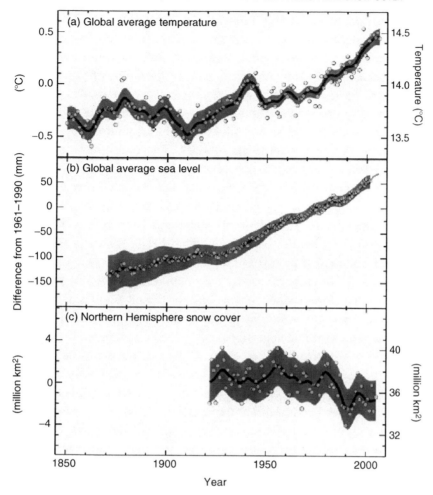

Figure 12.1 Observed (a) global average surface temperature; (b) global average sea level rise and (c) Northern Hemisphere snow cover for March–April. All changes are relative to corresponding averages for the period 1961–1990. The shaded areas represent decadal averaged values and their uncertainty, while circles show yearly averages (IPCC, 2007a).

increase in the global radiative forcing of about 10%, from about 2.6 to about 3.0 W m^{-2} (see (IPCC, 2007a)).

The global mean temperature did not increase between about 1940 and 1970. Since then, however, the increase has been about 0.4–0.5 °C and is undoubtedly primarily due to the enhanced greenhouse effect. No effort to simulate the global temperature changes during the twentieth century with the aid of global climate models as a result of a combination of natural and human causes has been successful unless it also took an enhanced greenhouse effect into account.

The direct warming due to human-induced greenhouse gas concentrations is rather modest. Doubling of the atmospheric carbon dioxide concentration is determined to yield a warming of only about 1.2 °C, but this in turn leads to an increase of atmospheric water vapour, which is a strong greenhouse gas. Such an increase has been confirmed by observations. *The sensitivity of the climate system to a doubling of carbon dioxide concentrations is therefore assessed rather to be about 2.5 °C with an uncertainty range of 1.5–4.5 °C.* This quite robust feature of the climate system has been derived with the aid of global climate models checked against observations from about the last 40 years.[4] There are still some who express doubts about the conclusion that the observed global warming during the twentieth century is exceptional and reject the view that the observed changes during the last 40 years have primarily been caused by enhanced greenhouse gas concentrations in the atmosphere. This view still sometimes finds its way into media, but has no support in the scientific literature and has been refuted by the IPCC.

There are also a number of other observations that support the conclusion that the ongoing change is primarily caused by human activities. In particular, the temperature in the lower stratosphere has decreased significantly, which is expected in the case of a warming of the lower atmosphere as a result of the enhanced carbon dioxide concentrations. Also, the daily temperature range at the earth's surface has decreased over parts of the continents, which is similarly consistent with the physics of an enhanced greenhouse gas warming.

The global warming has occurred in spite of the fact that global air pollution (aerosols) has increased considerably during the last two–three decades, not least as a result of emissions in developing countries. On average, aerosols cool the atmosphere by increasing the reflection of solar radiation back into space. It is therefore likely that the global warming due to changing greenhouse gas concentrations in the atmosphere has been reduced, at present by perhaps 25–40%, and probably even more regionally. This implies that if measures were taken in the future to reduce air pollution and smog in industrial regions for health reasons, the enhanced greenhouse effect would became more prominent.[5] Also, the response of the climate system to increased greenhouse gas concentrations is delayed because energy is needed to warm the uppermost layers of the oceans. The combined effect of the cooling due to the presence of more aerosols in the atmosphere, and this inertia of the climate system implies that we today might be seeing just about half of the ultimate warming due to the increase in greenhouse gas concentrations in the atmosphere that has occurred so far.[6]

The global hydrological cycle has intensified, which is expected in a warmer world with enhanced amounts of water vapour in the atmosphere and increased rates of precipitation and evaporation. Precipitation has also on average become somewhat more intense in some continental regions, while more dry spells have occurred in the subtropics.[7] The extent of the sea ice in the Arctic is decreasing and the ice cover is becoming thinner. Increased melting has been observed at lower levels of the Greenland ice sheet (see Figure 12.1(c)). The extent of mountain glaciers has declined markedly all over the world.[8]

Sea level has risen about 18 cm during the twentieth century and is expected to continue to rise at an increasing rate, which undoubtedly is the result of the melting of glaciers on land and the expansion of sea water as temperature rises (see Figure 12.1(b)). Coastal regions are becoming more vulnerable to storms, such as those that in recent years have hit the tropics, e.g. Bangladesh, India and the coral islands.

The transfer of carbon dioxide from the atmosphere to the oceans and also into the terrestrial systems, i.e. forests and soils, is slow, and about 45% of the human-induced annual emissions are accumulating in the atmosphere, in spite of the fact that these other reservoirs are huge. As the ocean surface water warms the solubility of carbon dioxide decreases and the vertical stability of ocean waters increases, which in some regions reduces the transfer of excess carbon dioxide to deeper layers, away from the atmosphere.

The IPCC has developed a wide range of scenarios of future emission of greenhouse gases and aerosols and assessed possible increases in their concentrations. The difference between future carbon dioxide concentrations for the most expansive and the most constrained projections of emissions will barely be noticeable until a decade or two has gone by (see Figure 12.2). Even though such scenarios depend on the assumptions made regarding future changes in global society, about which we merely can speculate, the range of the global mean temperature increase by 2100 above preindustrial values is projected to be between about 2 °C and about 6 °C (see Figure 11.1). Even the central value of this range would imply a substantial change of the rather stable climate that has implicitly been assumed so far in human plans for the future.

A very substantial and sustained decrease of emissions of carbon dioxide as well as human emissions of other greenhouse gases is required in order to achieve stabilisation of the ensuing greenhouse gas concentrations in the atmosphere and a slow-down of the ongoing change of the global climate. *The rate of the present increase of global emissions from burning fossil fuels is increasing rather than decreasing because of the rapid industrialisation* in developing countries, particularly in China and India. Even the modest goals agreed in Kyoto of a reduction of emissions by developed countries, which would have been just the first step towards stopping global warming probably will not be achieved.

The inertia of the human response depends on the reluctance of people and countries to act, as well as on the time that is required to change the complex infrastructure of the modern society, and especially to develop, plan and build new energy supply systems. It will therefore most probably be several decades

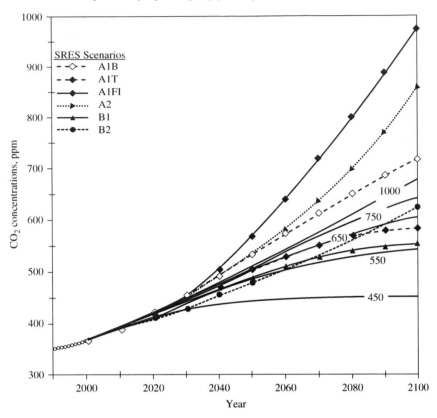

Figure 12.2   The observed average annual concentrations of atmospheric concentrations of carbon dioxide, as measured at the Mauna Loa (Hawaii) and the South Pole observatories from 1990 to 1999 are shown by small circles. Specified concentration pathways leading to stable concentrations ranging from 450 to 1000 ppmv are shown by the labelled solid lines. Curves with symbols show concentrations from 2000 to 2100 resulting from six specified *SRES* emission scenarios (Bolin and Kheshgi, 2001).

before *stabilisation of emissions* of greenhouse gases to the atmosphere is possible. *Greenhouse gas concentrations* would not then stabilise until after the middle of the century and *global warming will still continue for decades* thereafter as a result of the inertia of the human as well as the natural systems (see Chapter 13). The twenty-first century will therefore gradually become characterised by a global climate that is quite different from what humankind has experienced before, the more so the longer it takes to impose more stringent restrictions on net emissions of greenhouse gases.

Thus, it becomes most important to keep the changes that have begun within 'reasonable' bounds, even though we do not know either what is adequate in this regard, or the magnitude of the efforts required to do so. We note, however, that the present total human-induced greenhouse effect is equivalent to a carbon dioxide concentration in the atmosphere that is well above half way towards a doubling of the equivalent atmospheric carbon dioxide concentration which implies an increase in the mean global equilibrium temperature of $1.5–4.5\,^\circ\mathrm{C}$. Obviously major efforts are required in order not to exceed this level of forcing due to enhanced atmospheric greenhouse gas concentrations. How quickly will future changes come about? What can be done? These are the core issues to be discussed in Chapter 13.

12.2.3 Regional and local changes

Global warming will also be characterised by regional changes, i.e. modifications of the major global atmospheric circulation patterns bringing about changes in the distribution of wet and dry climatic zones and ecosystems, ice conditions in the Arctic Sea, a slowly melting of the Greenland ice sheet, etc. Most of these are irreversible as seen from a human perspective, i.e. the climate would not return to the state that prevailed before human global impacts became significant, even if the greenhouse gas concentrations slowly returned to pre-industrial levels. This would in any case take centuries and markedly influence the global system in the mean time.

We as individuals will primarily be recording climate change as changes of the weather, and particularly so if such changes occur in our more immediate neighbourhood. It then becomes important to clarify what we can say with some confidence about the characteristics of local and regional changes, i.e. changes of the weather because of a future global climate change as compared with its past variability. I wish here to distinguish between three kinds of changes.

A warmer atmosphere will contain more water vapour, which is a major energy resource for the creation of extreme weather and may therefore intensify precipitation, storms and hurricanes, etc. Extreme weather events of this kind might thus become more frequent than before but not necessarily more destructive, To what extent? How quickly? Where? These are key questions that largely remain unanswered and they will probably remain so for some time still. Since extreme events by definition are comparatively rare phenomena, it is difficult to ascertain statistically on the basis of weather in the recent past to what extent changes of this kind are already on their way.

The possible change in the occurrence of tropical hurricanes is an interesting example. Recent analyses seem to indicate that the frequency and possibly also the intensity of tropical hurricanes off the southeastern and southern coasts of the USA are on the rise. The observed increase in the surface water temperature in the Gulf of Mexico, presumably a result of global warming, might be of significance in this context. Similar trends can be seen elsewhere in the hurricane-prone regions over the globe, although their significance is still being discussed.[9]

It is, however, also important to add that the serious damage caused by hurricane Katrina in 2005 was undoubtedly to a very considerable degree due to human exploitation of the coastal region around New Orleans without adequate protective measures having been taken against flooding. The protective levies were not maintained adequately in spite of the fact that the land was continuously sinking because of lowering the ground water level and changes in the water courses in this sensitive delta region. Vulnerability had increased considerably because of human neglect. In addition the hurricane struck New Orleans in a most unfortunate manner. The catastrophe must not be seen primarily as a consequence of global warming and illustrates well the characteristics of a chaotic system. Serious destruction is caused by slow changes of the general setting and accidentally occurring extreme events. The media often claim that storms and floods are already more common than during the twentieth century and have become more intense, indirectly implying that they might become even more severe in the future. But this is as yet difficult to conclude generally. Statistical analyses of where and to what extent such changes might be expected, particularly during the next decades, would be most valuable. Precise answers are, however, not likely to be obtained quickly because of the random nature of extreme events and the enhanced spatial resolution of the climate models that are required to deal adequately with issues on regional to local scales. However, the risk for unusual weather events has undoubtedly increased.

It has since long been recognised that *the global circulation of the atmosphere may have several preferred states* that occur with varying probabilities. Studies of the likely future changes in the general circulation of the atmosphere in a warmer world are designed to explore to what extent this is so and whether global warming will have a significant effect. It is in this context interesting to note the extended periods of unusually hot and dry weather that have occurred in some regions in recent decades, for example in parts of central and south Europe, and particularly in France and Portugal, in the summer of 2003. The pattern of excessive temperatures during that summer shows clear similarities with regional model projections of what might happen more frequently in a warmer world (Figure 12.3). In France in 2003 there were 12 000–15 000 more deaths, particularly amongst the elderly, than expected in an average year and about three times that many in the central Europe as a whole.[10] The comparatively very warm weather that prevailed for several months in large parts of Europe in the autumn and early winter of 2006–7 might have resulted from a similarly locked-in situation of the general circulation pattern for the area concerned. We simply do not yet know. Similar situations with severe droughts have been

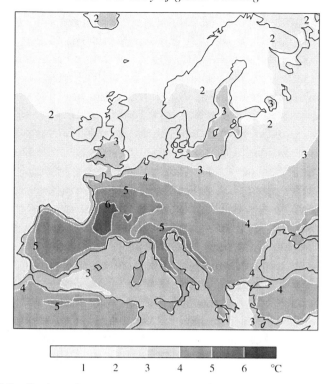

Figure 12.3   Projected average temperature changes for the summer (June–August) over Europe from 1961–1990 to 2071–2100 derived on the basis of the IPCC Scenario B2, making use of a regional climate model (Bolin 2004; courtesy of the Rossby Centre, SMHI, Sweden).

observed elsewhere, particularly in Africa and Australia, but are less well documented. It is still difficult to judge whether, and if so to what extent, these extreme situations result from the ongoing change of the global climate, but it is possible that they do. Their severity and frequency of occurrence in the future remain uncertain. Model simulations of regional climate changes in the subtropics indicate that more extended periods of hot and dry weather may become more common pole-wards of the dry subtropics and therefore may hit regions with large populations (e.g. the Mediterranean, the Middle East, India, Australia, South Africa and parts of southern USA)[11]. Admittedly, the likelihood of this occurring has not been well established. However, more extended periods of hot and dry weather seem in any case to be the most threatening feature of a warmer world at these latitudes. Securing adequate water supplies for a growing population is already an increasing concern, especially in Africa.

A third feature of significant regional changes of the climate system might be brought about by changing *interactions between the atmosphere and other components of the climate system, i.e. the oceans, ice sheets, sea ice and terrestrial systems.* For example, changes of the so-called thermo-haline circulation of the oceans, especially in the North Atlantic region might occur and be associated with a gradual decline in the intensity of the Gulf Stream (see Section 7.2). A most remarkable incident about 10 000 years ago was the rapid return to a cold climate in the region that then lasted for about 1000 years (the Younger Dryas), until the warming towards the present interglacial period was resumed. It seems plausible that the melt waters from the extended ice sheets at the time decreased the salinity of the surface waters in the North Atlantic region.

The Greenland ice sheet is at present melting more quickly along its rim than was the case just some decades ago, but on the other hand snow still seems to be accumulating over the ice sheet plateau.[12] Seen from a human perspective the melting of the Greenland ice sheet is a slow process. But at some time, probably more than centuries into the future, a situation of no return may be reached and the ice sheet might disappear in a matter of a millenium or more. The volume of the ice sheet as a whole corresponds to a sea level rise of about 6 m. The sea ice in the North Polar Sea is also melting and precipitation at northerly latitudes increasing. This all implies that the salinity of surface waters at northerly latitudes is probably decreasing. Whether this will weaken the Gulf Stream in a similar manner to 10 000 years ago and, if so, when this might occur is not possible to tell at present. There are, however, indications that the strength of the deep return flow towards the south is decreasing, though only slowly in the North Atlantic region. Changes in the extent of sea ice in the North Polar Sea and particularly its possible disappearance in the future in extended regions during the late summer season might have far-reaching consequences in neighbouring regions (ACIA, 2004). The likelihood of such changes and when they may occur cannot be foreseen with much certainty.

The El Niño phenomenon is characterised by close semi-periodic interactions between the ocean and the atmosphere in the equatorial regions of the Pacific Ocean. Its features have changed significantly during the last few decades. This has influenced the quasi-regular variations between the two opposite phenomena, El Niño and La Niña, and marked changes in the distribution of precipitation within the Pacific equatorial belt have been observed, This, in turn, has been associated with changes in the weather along the west coasts of both North and South America (and possibly even further eastwards) that have had significant socio-economic consequences in some of the areas that have been hit. Are these changes the result of the global warming? We do not know for sure but model analyses of the phenomena indicate that there might be a connection.

The message of this section is obviously a mixed one because of the difficulty of pinpointing more precisely what is going to happen on regional and local scales, but the list above is certainly not exhaustive. And as has also been constantly emphasised, surprises cannot be excluded. What then is the answer to questions like: What might we expect as individuals in a warmer world? How serious might a global change of climate be and how urgent is it to act? First and foremost it is important to bring home the following.

It will not be possible even with substantially greater research efforts to project in more detail, and with more certainty how climate and weather might change in a long-term perspective on regional and local scales and we cannot rule out surprises. It is therefore very important to view the future in terms of increasing risks for events of this kind to occur. Further research efforts are of course still important, but we do not know how reliable more detailed predictions will be until it has been possible to validate the research on the basis of new data, which can only be gathered slowly, i.e. in pace with the ongoing climate change. We may identify areas as being more vulnerable, but the key message must be: a significant change of the earth's climate is on its way. Even if more far-reaching mitigation efforts are initiated soon, it will take many decades, probably more than half a century to stop the ongoing climate change, and even longer if more intensive mitigation initiatives are delayed. Early and sustained preventive action is needed in order to keep expected changes within bounds.

Nevertheless, as mentioned before, there are objections to the IPCC conclusions and thus also to a summary of the kind that I have attempted above. These objections are, however, seldom found in the peer-reviewed scientific literature, but rather in personal interviews and, of course, on the internet. Home pages expressing doubts are quite numerous, but are often simply not trustworthy. However, this 'grey' literature sometimes catches the attention of a wide circle of non-specialists and is misleading the general public.

On the other hand, interpretations of the ongoing changes in weather and climate also feature in the daily press and the climate change threat is often illustrated by rather spectacular and exaggerated stories. One reason for this is the simple fact that everyone knows what weather is and we all experience its variations. It is then too easy to paint a vivid picture that shows a selective and spectacular version of what the future might hold. This undoubtedly complicates the handling of the climate change issue. Nevertheless, simple factual stories with proper interpretation are in the long run what counts.

It is a simple and regrettable fact that the delay in the Kyoto Protocol coming into force means that stabilising the climate has been significantly delayed and stabilisation at a rather low level of atmospheric greenhouse gas concentrations has become much more difficult and costly to attain. Huge investments have been made in the course of the last decade in expansions of the global energy supply system primarily based on the use of fossil fuels, and particularly coal (global emissions of carbon dioxide have increased by about 20%), that for economical reasons are not easily abandoned. This means that the world has become further locked into a fossil fuel dominated energy supply system.

12.3 Impacts and adaptation

Human-induced climate changes have already occurred and significant further changes cannot be avoided, but Nature will not collapse in a global catastrophe, though it will change markedly. The major threat will come from the difficulty that global society will have in adapting to these rapid major changes of the environment and particularly their implications for our ability to establish a sustainable future for a still growing world population. The rate of change of the global climate needs to be kept within bounds by worldwide cooperative efforts of mitigation, and adaptation to unavoidable changes is obviously also essential. The implications must necessarily be explored in terms of national efforts because of the need to assess local vulnerabilities. However, developing countries will be more severely hit and have less capability to take the necessary protective measures. The developing countries will need assistance from the industrialised countries to deal with their geopolitical situation. While a number of industrialised countries have made substantial efforts to determine their vulnerability, few developing countries have. The Climate Convention has as yet taken few steps to improve this situation. It is important to translate the global climate change assessments of the expected range of future changes into specific programmes; attention needs to be given to reliable risk analyses for the likely gradual regional and local changes of climate and their implications for traditional societal activities and priorities.

For example, what is the effect of an increase of the surface ocean temperatures on coastal ecosystems and does this in turn affect the livelihood of people in these regions? What effect does a hotter climate (e.g. in subtropical countries) have on the water supply? How does the melting of the permafrost threaten the infrastructure in the tundra region? What needs to be done to protect the inhabitants of flood-prone areas, where intense precipitation events might become more devastating? How sensitive are the natural habitats that would be stricken by excessive flooding, e.g. shorelines or river valleys? How will the increasing probability of hot spells with drought influence agriculture in the subtropics? Do changes in the growing season mean that different strains should be grown in agriculture and forestry? How sensitive are local communities to changes of frequency and intensity of extreme and destructive weather events and what are the implications with regard to available protective measures? Where might there be an increasing risk of storms and gales, and how will this influence the insurance industry?

If efforts of this kind have not yet begun, there are indeed good reasons to start soon. This is also the kind of activity for which the Global Environmental Facility at the World Bank carries the responsibility to support developing countries financially.

Careful studies of past interannual variations of climate and weather and their impacts are the first steps towards answering questions of this kind and are beginning in several countries. They will serve a dual purpose, i.e. not only supply a basis on which desirable protective measures can be taken, but also bring about a greater awareness among people in general about the possible implications of climate change.

The climate changes that have occurred so far, and those that are unavoidable in the near future, will be with us for many decades or centuries because of the inertia of the climate system. Gradual adaptation is therefore essential in order to limit the impacts of climate change as far as possible.

12.4 Science, media and the general public

There has been an unfortunate polarisation of the way the media report the climate change issue. There has been, and still is, focus on either environmentalists who paint vivid pictures of catastrophes or sceptics who consider that the climate change issue is being much exaggerated. It is essential that the media describe in more realistic terms what might be expected to happen and what should be done about it. This is not easy, however, because few of those that are engaged in the science try to grasp the socio-economic complexities and are able to judge what is important in this regard. This results in very 'noisy' media messages.

However lucid a written description of the likely future change of the global climate might be, a proper understanding and engagement of the public cannot be achieved only by articles in the daily newspapers, magazines and books. Other media have a very important role to play. Film and television are obvious channels to be used in order to communicate with the general public.

This has, of course, already been done, but should not be left exclusively to commercial exploitation. Snapshots as presented in news programmes emphasise dangers and are often polarised between 'for' and 'against' in a manner that distorts the message. An example of the distortion of reality is the production of films like 'The day after tomorrow' (2004) which has been shown worldwide. The scientific input was very meagre and an understanding of the issue at stake totally lacking. The producer excelled in making use of the most advanced technical facilities to create a drama that has very little to do with the real environmental problem that we are facing. The human suffering arising from a serious climate change is its central theme but the film is far-fetched and shows a naivety beyond belief. Of course, the film was not intended to be a serious documentary but if it has had an impact at all it is to create fear of or indifference to climate change, neither of which is very helpful in attempts to deal with the issue.

In 2005–6, on the other hand, two interesting efforts were made to reach out to the broader public. With support from Swedish Television, a group of free-lancers[13] produced the documentary 'The planet'. This was done in cooperation with the International Geosphere Biosphere Program (IGBP).[14] A number of leading researchers in the field, chosen from around the world, appear in the film and a central role is played by Dr Will Steffen, executive director of IGBP until 2005. The climate change issue is seen as part of a broad exposé of the increasing exploitation of the natural resources. The issues of sustainable development and ongoing changes are shown in vivid pictures and bring home very well what happened during the twentieth century and the changes that may be on the way, not least global warming and its ramifications. Because of the close association with the scientific community the facts are largely correct and balanced and well presented. The film is, however, in a way somewhat anonymous. The spectacular photography becomes monotonous and viewers have difficulty associating themselves with the people that appear on the screen, even though there are some scenes that succeed well in this regard. The concluding sequences are also valuable in showing that it is difficult for many to accept that such a worrisome future lies ahead. This indeed illustrates the necessity of becoming personally touched and thereby engaged. Psychologically it is difficult to absorb and accept facts from many sources if we are not also given examples of measures that can help us out of the quandary in which we find ourselves.

The other film that appeared in 2005–6, 'An inconvenient truth', produced by D. Guggenheim, shows the former US Vice-President Al Gore lecturing to an audience about the global warming issue against a background of what is happening around the globe. His story is, however, not always adequaely founded in the basic scientific knowledge that is available. For example, the changes of carbon dioxide and the global mean temperatures during the last 650 000 years are presented in a manner that implies that the ice ages were caused by variations of atmospheric carbon dioxide, which of course is wrong. Nevertheless, the spectacular way in which the graph is presented as a sky-rocketing curve of the ongoing carbon dioxide increases is impressive and will probably be remembered by many. This graphic and some other exaggerations can be interpreted as symbolic statements that we are entering an era which may be very different from that to which we are accustomed. On the other hand, the programme says that the expected heating around the North Pole might be $+12\,°C$, an extreme that misleads the viewer about what might happen in the Arctic, and Al Gore does not inform us how quickly or slowly an altered future may come upon us.

However, his openness about tragedies in his private life, his honesty and personal engagement in the global warming issue, and not the least his rhetoric make him seem very credible. His presentation is not just about facts and numbers, but he rather emphasises that we all have a moral obligation to take care of our only earth.

What is lacking in both films, however, is an emphasis on how fast things might be changing, what we should be doing, the difficulties we will encounter when trying and what opportunities there are to take precautionary measures. A simple statement that we have to use renewable energy is next to meaningless unless some idea about what this may imply is also given. We also need to be much more careful about the way we use energy in our daily life. As we shall see in the final chapter, these aspects of the issue are by no means easy to grasp, but are essential to penetrate. The energy issues that are now becoming acute for all of us mean that the climate change will also have to be brought more to the forefront.

13

Climate change and a future sustainable global energy supply

Combating climate change implies resolving the controversial issue of how to renew the global energy supply system, and simultaneously reduce emissions of greenhouse gases to the atmosphere.

13.1 Delayed action in spite of trustworthy scientific assessments

Mitigating climate change is finding ways and means to reduce emissions to the atmosphere and to enhance sinks for the atmospheric components that disturb the radiative balance of the earth with space. But it is equally important that this should be done in a way that minimises costs and the degree of disturbance to countries and people, ensures a sustainable development and is in this sense politically acceptable. The issue is thus not primarily a technical and economical one, but societal and political, even though technical and economic analyses of the means to deal with it are very important as a basis for identifying and reaching agreements on action plans and how they are to be implemented. A brief overview of matters that are of prime concern in this context will be given in this concluding chapter, especially with regard to the basic scientific and technical knowledge about the global society that will be required. The best common understanding of this issue is a prerequisite in order to mitigate a human-induced climate change quickly enough. This is, however, not the place to analyse the technically complex issue of alternative future global energy systems in detail.

Stabilising the global climate must necessarily be a worldwide cooperative undertaking, but it is important to recognise countries' differing ability to contribute. This is well expressed in the Climate Convention.

Article 3
    1 . . . the Parties should protect the climate system for the benefit of present and future generations of humankind, on the basis of equity and in accordance with their

214

common but differentiated responsibilities and respective capabilities. Accordingly, developed countries should take the lead in combating climate change and the adverse effects thereof . . .

Article 4

2 The developed countries . . . commit themselves specifically as provided in the following:

a. Each of these Parties shall adopt national policies and take corresponding measures on mitigation of climate change, by limiting its anthropogenic emissions of greenhouse gases and protecting and enhancing its greenhouse gas sinks and reservoirs. These policies and measures will demonstrate that developed countries are taking the lead in modifying longer-term trends in anthropogenic emissions . . .

These principles are also reflected in the Kyoto Protocol and have been elaborated in the Marrakech Accord reached at the eighth conference of the parties to the Climate Convention in 2002. However, efforts to stabilise greenhouse gas concentrations in the atmosphere, and thereby the climate, will probably have barely begun by the end of the first commitment period of the Kyoto Protocol in 2012, 20 years after the Climate Convention was opened for signatures in Rio 1992. It does not seem likely that even the modest goals of the Kyoto Protocol will then have been reached. This slow pace is indeed worrisome. Two things are particularly important in this context. We must better appreciate the scope of the problem ahead of us and we must get a general idea of what can be done, i.e. opportunities and constraints.

13.2 Past and future emissions of greenhouse gases and aerosols

13.2.1 Carbon dioxide

As already stated in Chapter 12, the atmospheric carbon dioxide concentration rose to about 380 ppmv in 2006, i.e. an increase as a result of human emissions since preindustrial times of about 36%, with the rate of increase being about 2 ppmv per year and still increasing. About 330 Gt C (i.e. about 1250 Gt of carbon dioxide) have been emitted into the atmosphere since the middle of the nineteenth century due to fossil fuel burning and cement production.[1] In addition, a net total of about 130 Gt C has been emitted because of deforestation and changing land use (regrowth as well as forestry activities in general have been considered in deducing this number), i.e. the net human-induced emissions total about 460 Gt C[2]. The carbon dioxide content of the atmosphere increased during this time by about 210 Gt C, i.e. about 45% of total emissions have stayed in the atmosphere, while about 250 Gt C have been taken up by the oceans and the terrestrial ecosystems, the latter partly because of more intense photosynthesis in an atmosphere enriched in carbon dioxide. The average annual distribution of

Table 13.1.   *Average annual carbon budget for the*
atmosphere in the 1990s in Gt C per year

Burning of fossil fuels	6.3 ± 0.4
Storage in the atmosphere	3.3 ± 0.1
Ocean uptake	1.7 ± 0.5
Net terrestrial uptake	1.3 ± 0.7
Emissions from land use change	1.6 ± 0.8
Gross terrestrial uptake	2.9 ± 1.1

net transfers between major carbon reservoirs during the 1990s are given in Table 13.1.[3] However, these values had changed considerably by 2006. The use of fossil fuels and cement production now yields emissions of about 8.0 Gt C per year and the total emissions, including emissions because of deforestation and changing land use well above 9.5 Gt C per year, the annual increase of the atmospheric carbon dioxide concentration is about 3.8 Gt C per year, i.e. approaching 2 ppmv. The oceanic uptake is presumably also increasing somewhat because of the continued increase in the atmospheric concentration. Similarly the net terrestrial uptake has been further enhanced due to increased photosynthesis. A more favourable climate at northerly latitudes may also have contributed to the carbon uptake. This uptake by the terrestrial systems seems generally to have increased during the last few decades of the twentieth century, which has contributed significantly to modulate the increase of the atmospheric carbon dioxide concentration. However, these simple extrapolations to today do not tell us much about what long-term changes to expect.

Stabilising the climate will necessarily require limitations of future emissions and where possible enhanced sink mechanisms in the oceans as well as on land.[4] The interpretation of the wording in the Climate Convention '. . . The ultimate objective of this Convention is . . . to achieve . . . stabilisation of greenhouse gas concentrations in the atmosphere at a level that would prevent dangerous anthropogenic interference with the climate system . . .' must, however, be a matter for political judgement.

The global carbon cycle is reasonably well understood and can be used to assess the 'permissible' emissions in order not to exceed concentrations at some prescribed levels. As discussed in Section 8.3, a number of concentration levels for stabilisation have been explored, i.e.. 450, 550, 650, 750 and 1000 ppmv, in order to illustrate different choices that might be of political interest.[5] The outcome of these analyses, which were carried out in mid-1990s and which are shown in Figure 8.1, is still largely valid but remains rather uncertain. Several assumptions must be made in order to arrive at a quantitative assessment. Above all, the likely feedback processes due to probable changes of the climate during the twenty-first century may influence the uptake of carbon dioxide by the oceans as well as the terrestrial systems, but have been dealt with rather schematically because of insufficient knowledge.[6] We are not yet very sure how the different feedback mechanisms will play out. Nevertheless, in the efforts to achieve sustainable development this set of scenarios provides important information about the limitations of future emissions imposed by the natural system.

To simulate two alternatives of the socio-economic inertia the emissions of two sets of scenarios were shown in Figure 8.1, in one they have been assumed initially to continue to increase for some time, while the slow-down of the emissions begins more quickly in the other set. The accumulated 'permissible' total net emissions during the years 1990–2100 in the case of stabilising at 450 and 550 ppmv were assessed to be 640–800 Gt C and 880–1060 Gt C respectively. About 120 Gt C has already been emitted during the 16 years since 1990. Thus net average annual emissions of not more than 5.5–7.2 and 8.0–9.9 Gt C, respectively, may be emitted during the remainder of this century including the net exchange with the terrestrial system due to sources and sinks, if stabilisation is to be achieved at 450 and 550 ppmv. Present emissions due to fossil fuels use have already exceeded these limits or will soon do so, since they are continuing to increase. The natural uptake of the terrestrial biosphere has, in spite of the ongoing deforestation, modulated the rate of increase of atmospheric concentrations. In the long run, however, this counter balancing process cannot keep pace with a continuing increase of emissions due to fossil fuel burning, but must begin to decline well before the middle of the century to permit either of the two different stabilisation levels to be realised.

We need, however, urgently to focus on the very different situations of the developed (industrialised) and developing countries.[7] Figure 13.1 shows the average annual per capita emissions of carbon dioxide due to fossil fuel burning in 1999 in nine regions of the world. Note first of all that the differences between the regions were huge and they still are. The global, annual, average, per capita emissions from developed countries including countries in economic transition (i.e. the Russian Federation and other former states of the USSR) were on average about 3.0 ton C. The USA, Australia and Luxembourg had the highest emissions, more than 5 ton C per capita, and none of the industrial countries emitted less than about 1.6 ton C per capita. South Korea has rapidly increased its energy use, developed a very competitive industrial sector and in that sense joined the OECD countries of the west with an average emission of about 3 ton C per capita. Developing countries in general, however, emitted on average only about 0.6 ton C per capita (a few of the oil-producing countries, however, 5–8 times more). The inequity with regard to energy use, and thus the use of fossil fuels, between the countries of the world is a striking sign of the global

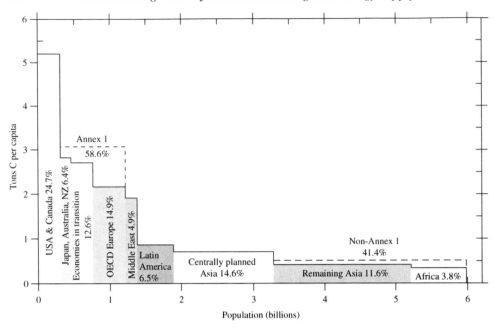

Figure 13.1   Per capita fossil fuel emissions in 1999 averaged for nine geo-graphical regions and grouped into Annex-1 (industrial) and Non-Annex-1 (developing) countries. The height of the bars gives the average per capita annual emissions of each region. The width of the bars gives the population. The area of a bar is thus proportional to the 1999 carbon dioxide emissions from fossil fuels and cement production (Bolin and Kheshgi, 2001).

differences in socio-economic development that still prevailed at the beginning of the twenty-first century.[8]

However, this situation is changing quite rapidly. The emissions in several industrialised countries are still on average increasing by up to about 1% per year. The USA has not ratified the Kyoto agreement and its emissions have increased by about 16% during the 12 years from 1990 until 2002, while Japan and the EU have increased their emissions by 14 and 3% respectively.[9] As compared with the Kyoto agreement the carbon dioxide emissions in 2002 from the USA, Japan and the EU were 23%, 20% and 11% above their respective assignments. It is not likely that any of them will be able to meet the goals set for the first commitment period, 2008–2012, as given in the Kyoto Protocol, although the EU has been more successful in not increasing emissions much above the 1990 level. The Russian Federation and the Ukraine are now again increasing their emissions after a drop by more than 30% during the 1990s, but have not yet reached their emission levels in 1990, which was their limit according to the Kyoto Protocol. Whether they will exceed their given quota is still unclear, but it seems likely that they will not.

The economies of developing countries have grown faster than envisaged a decade ago and their use of fossil fuels has increased on average by 40–60% since 1990. Within just a few years their total carbon dioxide emissions will exceed emissions from industrialised countries, but their per capita emissions will on average still be only about 0.80 t C per year. However, emissions from China are increasing very quickly due to its very rapid industrialisation and are now approaching 1.0 t C per year, while India is similarly expanding its use of fossil fuels, although more slowly, and has started from a much lower level. On the other hand, economic development in most African countries is lagging behind and their emissions are increasing more slowly.

As already mentioned global emissions due to the burning of fossil fuels and cement production (now about 8.0 Gt C per year) have reached an all time high. This also means that the assessment of total emissions in 2010, which based on the assumption of adherence to the Kyoto Protocol was deduced in 1998 to be 7.1 Gt C per year, is no longer valid. The same is the case for my estimate in 2001 of 7.6 Gt C per year.[10] In fact the rate of increase of global carbon dioxide emissions has been accelerating in recent years, while the percentage contribution to the world energy supply from, for example, the use of nuclear energy and hydropower has decreased because of the comparably modest investments that have been made. Measures undertaken so far to slow down the increase of global net emissions due to fossil fuel burning have been very inadequate. Almost a decade has been lost in trying to deal with the ongoing accumulation of carbon dioxide in the atmosphere. Note, however, that the global emissions due to changing land use may even have been reduced, as the *SRES* study also assumed might happen, but the future development much depends on the development of policies for the exploitation of tropical forests. The statistics from East and South-East Asia are very uncertain because of extensive illegal logging. In summary, the total annual carbon dioxide emissions are today well above the stabilisation curves shown in Figure 8.1 and they are still rising.

The decision at Kyoto that industrialised countries should aim by 2008–2012 to reduce their greenhouse gas emissions by about 5%, was not based on thorough analyses but was rather an ad-hoc decision that was tentatively agreed in order to take the first steps towards mitigating global climate change.[11] In spite of its modest goals the Kyoto Protocol did not come into force until 2005 and the possibility of reaching these targets was much reduced. The Climate Convention negotiations must explore what can now be achieved during a second commitment period to make up for this serious delay.

Of course, we cannot predict the future of humankind, so we cannot predict the future increase in greenhouse gas concentrations in the atmosphere beyond the few decades in which projections can be based on the inertia of the global socio-economic system. The set of base scenarios (*SRES*) that the IPCC completed in 2000 were assumed not to be influenced by explicit political choices but rather were aimed at exploring the implications of different possible pathways towards a future global society and the sensitivity of the climate system to various assumptions about the emergence of new sources of primary energy.[12]

The scenarios assumed that the world population would reach some alternative levels between 7 and 16 billion by the year 2100, that the gross domestic product would increase annually on average by about 2% but that economic development might be different in industrialised and developing countries, allowing the latter gradually to catch up with the developed countries. Some of the scenarios involve future extensive use of fossil fuels, others depict a gradual change to renewable sources for primary energy supply and the development of what might be seen as a more sustainable energy supply system. Altogether some 40 scenarios were produced by six research groups and four basic types were analysed more closely (see IPCC (2000b)).

Levels of carbon dioxide and other greenhouse gases as well as air pollution in general were estimated and the changes in their concentrations in the atmosphere deduced. Assessments of future carbon dioxide concentrations were made using available carbon cycle models (see Figure 12.2). Note that all scenarios show a carbon dioxide concentration in 2100 from about 600 ppmv (SRES B1) to in the most extreme case (SRES A1F1) more than 900 ppmv. No scenario reached stabilisation. It is noteworthy that the world population in the scenario B1 with the lowest emissions during the next 100 years was assumed to reach a maximum of about 9 billion in the middle of the century and then to decline to about 7 billion by 2100, which of course partly explains the modest increase in this case.[13] These technical analyses might still serve as starting points for discussions about the impacts and damage due to human-induced climate change. Because of the rather large ranges of the key parameters that were allowed, it seems likely that the future greenhouse gas concentrations will stay within the given range until about the middle of the century or somewhat later if no preventive measures were taken.

In view of the very different per capita emissions between industrialised and developing countries in 2000, it is important to look at the different emission pathways that are followed by these two groups of countries in the scenario constructions, although only the gross features are reasonably trustworthy and therefore of interest. Figure 13.2 shows the *annual per capita emissions* for industrialised and developing countries separately during the twenty-first century for key scenarios and also the approximate per capita emissions that would be required in order to stabilise the global carbon dioxide concentrations at the alternative levels between 450 and 1000 ppmv, assuming a medium scenario for the population development. As seen from the figure, SRES B1 comes closest to

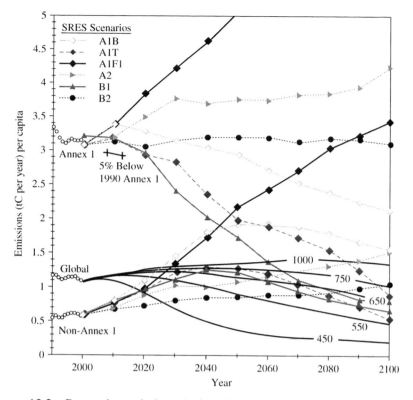

Figure 13.2   Per-capita emissions during the next 100 years. The black circles
show the per capita emissions from 1990 to 1999. The short line shows emissions
5% below the Annex-1 rate during the commitment period 2008–2012 as aimed
for in the Kyoto Protocol. Deduced fossil fuel emissions leading to pathways
towards stabilisation of atmospheric concentrations from 450 to 1000 ppmv are
shown by solid black lines. Per capita fossil fuel emissions for Annex-1 and Non-
Annex-1 countries according to six IPCC scenarios are shown by the curves and
symbols as detailed in the key. Since the different scenario projections assume
quite different changes in the global population, the curves are not more precisely comparable, but still give an overall understanding of the key issues at
stake politically. To reach stabilisation, the emissions pathways for Annex-1
countries (at present well above any of the stabilisation curves) and those of
Non-Annex-1 countries (at present consistently below) should asymptotically
and simultaneously approach the particular stabilisation curve aimed for if equity
with regard to carbon dioxide emission levels is to be achieved (Bolin and
Kheshgi, 2001).

stabilisation of the carbon dioxide concentrations at a level of about 600 ppmv for
both categories of countries and the emission paths do not reach above 1.3 t C per
capita. This implies an average reduction of the per capita emissions for industrialised countries during the century of about 70% and an average increase for
developing countries of their per-capita emission of about 50%, most of which
would have to be accomplished by the middle of the century. This confirms my

statements about emissions during the twenty-first century at my last sessions with the IPCC and at the Framework Convention in 1996 and 1997 (see Section 9.3.1). It should, however, be emphasised that the per capita emissions are just one parameter that can be used as a basis for burden sharing, but because of the striking differences of greenhouse gas emissions between industrialised and developing countries it is likely to be a very important one.

13.2.2 *Emissions of other greenhouse gases and aerosols*

It is important to recognise that about 35% of the enhanced greenhouse effect is brought about by increasing concentrations of a number of greenhouse gases other than carbon dioxide in the atmosphere, due to human activities and that aerosols also influence the radiative transfer through the atmosphere (see figure 11.2).

The reasons for the increasing *methane* concentration in the atmosphere are more diverse and difficult to assess. Leakage during the exploitation of natural gas, oil and coal resources may have been large until a couple of decades ago, but is better controlled today, partly for economic reasons. The diffuse emissions from coal mines are, however, difficult to stop. The increasing world population has meant a large increase of agriculture (particularly rice cultivation) and animal husbandry (there are about 1500 million cattle in the world today), which are very significant sources of methane. Nevertheless the atmospheric increase has slowed down during the last decade, though special efforts will be required to prevent further increases, because the world population will presumably continue to increase for decades to come and so will agriculture.

Note, however, that methane decomposes rather rapidly and has a mean residence time in the atmosphere of only about 10 years. If emissions are reduced, a corresponding reduction of the methane concentration will come about quite quickly. On the other hand, there is one major uncertainty. The permafrost regions in northern Canada and Russia and the bottom sediments in the Arctic Ocean contain huge amounts of methane clathrates (see Table 13.2). As mentioned previously, a warmer climate might therefore lead to increasing emissions of methane, but we do not know under which circumstances this might become a significant threat.

The gradually increasing emissions of *nitrous oxide* stem primarily from agriculture and forestry and are above all due to the increasing use of industrially produced fertilisers which increase the concentration of nitrogen compounds in the soil. The natural chemical transformations and circulation of nitrogen compounds in the soil (denitrification processes) release some of the nitrogen that is fixed in the industrial production of fertilisers back into the atmosphere in the form of nitrous oxide. An increasing use of bioenergy would also lead to enhanced emissions of nitrous oxide to the atmosphere. Its decomposition in the atmosphere is slow (the concentration would be reduced naturally by 50% in about 150 years if no further emissions occurred). More fertilisers will probably be used in the future and it is likely that atmospheric concentrations of nitrous oxide will continue to increase during most of the twenty-first century, although rather slowly. Emissions from the industrial sector are also playing a part but can be more easily controlled.

Enhanced *ozone* concentrations in the troposphere are primarily caused by atmospheric air pollution, while the decrease in the stratosphere is the result of the presence of *CFC gases*. Efforts to decrease air pollution and the use of CFC gases mean that changes of atmospheric ozone concentrations (decrease in the stratosphere and increase in the troposphere) might contribute less in the future to an enhancement of the greenhouse effect. The CFC gases themselves also increase the radiative forcing of the atmosphere. Even though emissions have decreased radically in the last 15 years, the CFC gases that are already in the atmosphere decompose only slowly and their concentrations will therefore probably decline slowly during the twenty-first century.

Because of the short residence time of particulate matter, i.e. *aerosols*, in the atmosphere their concentrations as well as their geographical distributions depend on the location of source regions, and vary largely in line with the emissions. While air pollution has decreased in developed countries (western Europe and North America), there have been marked increases elsewhere, particularly in densely populated regions, such as India, southeast Asia and China, and pollutants can nowadays be traced over major parts of the Indian and the Pacific Oceans, carried by the winds at upper levels. The rapidly increasing numbers of cargo ships that cross the oceans using the lowest grade of oil contribute quite significantly to the global increase in pollution. Most types of aerosols reduce the greenhouse effect ('dimming'), but the magnitude of their modification of the radiative forcing is not well known. However, emissions might decline in the future because of the threat they pose to public health, but we can only guess how quickly this might happen.

The radiative forcing due to human-induced gases and aerosols at the turn of the century is summarised in Figure 11.2. The total enhanced forcing has increased by 6–8% since then and is now about 3.0 W m^{-2}, primarily due to the increase in the carbon dioxide concentration (see IPCC (2007a)). It also seems likely that the relative importance of carbon dioxide will increase for some time, at least until decarbonisation of energy production comes into play effectively, while aerosol concentrations might well decrease soon.

13.3 Primary energy reserves and resources and their utilisation

It is important first to provide some idea about the primary energy resources that are available for use. The earth's crust contains stored energy that may be exploited; essentially fossil fuel and nuclear energy. These are obviously both finite. Renewable energy, on the other hand, stems directly from solar energy, of which small parts are stored temporarily and become available in the form of bioenergy, hydropower, wind energy, wave energy etc. Such an overview is lacking in the IPCC AR4 (IPCC, 2007c).

It is quite striking that so far inadequate attention has been given to global sustainability and the threat of climate change when strategies for resolving the problem of the world's future energy supply are being discussed. The short-term issues of possibly limited supplies of energy, particularly oil, have been considered by industrialists and politicians to be more acute, and sustainability has received less attention. Trustworthy information is needed about the opportunities and limitations in the development of the primary energy sources (see, for example, Bengtsson, 2006a,b).

13.3.1 Geopolitics of fossil fuels

Just a few decades ago the fossil fuel reserves and resources were considered to provide a guaranteed secure long-term supply of energy and thus also a peaceful development of global society. Even though warnings appeared during the early 1990s, it was not generally recognised until about the turn of the century that a shortage of conventional oil might not be that far away. Accordingly, this possibility was not considered much when the Kyoto Protocol on climate change was agreed in 1997.

I will first focus on reserves and resources of fossil fuels because of their likely continued use for many years. Table 13.2 gives a summary of global reserves, resources and additional occurrences of fossil fuels in terms of the potential carbon dioxide emissions if these are used as fuel to provide energy. The estimated reserves are reasonably trustworthy, while the resources listed are considerably more uncertain.[14]

In spite of the uncertainties, the table provides a useful overview for present purposes. Available reserves and estimated resources of conventional oil and gas would suffice to raise the carbon dioxide concentration well above 550 ppmv, and to at least three times the pre-industrial level if the reserves and resources of coal were also exploited.

Conventional *oil resources* are, however, not abundant and their use might peak within about a decade. Similarly, *natural gas* resources, although considerably

Table 13.2. *Global reserves, resources and additional occurrences of fossil fuels in terms of potential carbon dioxide emissions compared with past emissions as of 2005 (in Gt C): from IPCC (1996b) and extrapolated to 2005.*

	Emissions		Reserves identified	Resources estimated	Additional occurrences
	1860–2005	2005			
Oil					
conventional	103	3.1	100	70	
unconventional			130	130	>250
Natural gas					
conventional	43	1.4	80	220	>150
unconventional			100	300	330
Clathrates					>12 000
Coal	185	3.1	1050	2500	>4000
Total	331	7.6	1460	3220	>16 700

larger, may start to decline in use towards the middle of the twenty-first century because of their increasing exploitation as a substitute for conventional oil and because the energy demand will most likely increase at least as quickly as during the last few decades for quite some time to come. *Coal resources*, on the other hand, might well last into the twenty-second century. High oil prices have led to increasing use of coal, not least in China. The gradual decline of oil assets has stimulated the development of technologies to produce liquid fuels from coal. However, this would even further increase the carbon dioxide emissions per unit energy produced.

The challenge for politicians and stakeholders will be to develop a global energy supply system and simultaneously satisfy the requirement to slow down and ultimately stop climate change, i.e. gradually to abandon the use of fossil fuels for the supply of energy. The measures to achieve this must be acceptable to the general public; in particular it is necessary to provide energy for development in poor countries and at the minimum to maintain current standards of living in industrialised countries. These are the central issues for consideration and have obviously a clear and serious north–south dimension.

Even though a gradual change of the global energy supply system is a long-term concern, short-term issues will influence politics markedly. A few examples may illustrate this. More detailed analyses of such aspects of the issue are most desirable and are being pursued.

First we consider the decreasing availability of conventional oil. Table 13.2 shows that a total of about 170 Gt C (possibly less) of reserves and resources in the form of conventional oil may be available to be exploited. The past record of discoveries indicates that it is unlikely that new large oil fields will be found. This implies that within a decade about half of the total resources originally available (using today's technology for extraction) will have been used (in total some 135 Gt C). Some of the estimated resources might by then have been discovered and become available as reserves, but oil production will probably decline gradually, while demand and prices will rise further. It seems unlikely that countries with oil reserves would increase production substantially to meet increasing demands rather than save the oil for the future when it could be sold at a higher price.

The oil reserves and resources are, however, very unevenly distributed as can be seen from Table 13.3. The dominating role of the Middle East is striking. Many of the Russian oil reserves are also located in regions close to the Caspian Sea and in southwestern Siberia. The assets in Europe and the USA are very modest. Note that the *annual use of oil* in the three large industrialised regions, North America, Asia Pacific and the EU at present correspond to as much as 7–10% of their domestic reserves and resources. The industrialised countries would thus cover their present use of fossil fuels for only 10–15 years if the same use of oil as today is assumed and there was no reliance on imports.

This is obviously a major political issue in the short term. Industrialised countries are becoming increasingly dependent on foreign assets, particularly from suppliers in the Middle East, and the supply will accordingly become more monopolistic than today. OPEC is strengthening its already dominating position.

Table 13.3. *Reserves of conventional oil as distributed in six major regions of the earth as a percentage of the total (about 170 Gt C) and in absolute amounts (Gt C); the annual consumption (extrapolated to 2005) in these regions as a percentage of the total emissions (about 3.0 Gt C per year) and in absolute terms (Gt C per year)*

Region	Reserves in % of total	Reserves in Gt C	Consumption in % of total	Consumption in Gt C per year
North America	5.5	9	32.0	0.96
South & Central America	8.9	15	5.7	0.17
Europe, Russian Federation incl. Asia	9.2	16	23.7	0.71
Africa	8.9	15	3.3	0.10
Middle East	63.3	108	6.0	0.18
Asia Pacific	4.2	7	29.3	0.88
Total		170		3.0

Basic data from British Petroleum (2004).

World politics as played out at present in Iraq and its neighbouring countries should also be seen from this perspective. The future supply of conventional oil is already and will become even more a major geopolitical issue. Note also that unconventional resources of oil, e.g. tar sand and oil shale in Canada and Estonia for example, are being increasingly exploited, because the costs are no longer viewed as excessive. However, this is often environmentally very detrimental to the immediate surroundings of the fields that are being exploited and the emissions of carbon dioxide per unit of energy finally produced are also much larger than for conventional oil, since the extraction of the oil is energy intensive. Seen from a climate change perspective, this is a most undesirable development.

The patterns of demand will also change, particularly because of the increasing need for primary energy in the developing world, especially in China and India, the home of about 2400 million people, i.e. about half of the population in developing countries. The increasing cost of energy has led to a polarisation between rich and poor countries within the developing world and also between rich and poor people within these countries, obviously to the disadvantage of the poor. This may especially hit Africa, which is already more troubled by disease and poverty than other parts of the world.

The exploitation of *natural gas* began comparatively late but has risen rapidly in the last several decades. The total reserves and estimated resources are, as far as is known, larger than those of conventional oil and they are still being exploited less rapidly (see Table 13.2). It is not likely that their use will peak much before the middle of the twenty-first century. Reserves and resources of gas are, however, also unevenly distributed over the earth. About 50% are found in the Russian Federation and some of the other former Soviet states. The Middle East countries also possess substantial reserves and resources, while again quite limited resources are located in industrialised countries except for the Russian Federation. Europe is becoming increasingly dependent on Russian natural gas and the Russian Federation has signed a contract with China about delivery of gas for its continued industrialisation. The USA is negotiating with Norway in order to secure additional natural gas resources (from the North Polar Sea) for its domestic needs. Securing future energy supplies for industrialised countries in the form of gas is receiving increased attention.

Coal reserves and resources are much larger than those of conventional oil and natural gas, and they are more evenly distributed around the globe (see, however, the last paragraph of endnote 14). Large deposits are found in the USA, China, India, Australia, as well as in some European countries, which all have limited reserves of oil and natural gas. In fact, coal is at present the key primary energy source for electricity production in many countries (its share is globally about 45%), and is increasingly serving this purpose, not least in China. It should, however, also be recalled that for the production of one unit of electricity in a fossil-fuel-based power plant using coal, the emissions of carbon dioxide are about 40%, respectively 70%, larger than if using oil or natural gas. Anyhow, a change to coal seems to be under way in several countries that possess such deposits.

Finally, huge amounts of *methane hydrates (clathrates)* are buried in permafrost regions in Canada and Russia and in the bottom sediments of the North Polar Sea (see Table 13.2). They have not yet been much considered as a future energy source. It is not clear how they might be exploited safely so that the methane would not accidentally leak into the atmosphere.

It is also important in this context to recognise the risk that the melting of ice in the North Polar Sea and the thawing of the permafrost might lead to natural releases of methane from these deposits, which would reinforce the global warming, but it is not known under which circumstances this might become important, nor is it an imminent threat because the residence time of methane in the atmosphere is quite short. A substantial accumulation of methane in the atmosphere from leaking natural resources does not seem likely at present.

The most imminent danger might rather be the competition for the fossil fuel resources, a geopolitical situation that might lead to conflicts long before the dwindling resources of oil and later natural gas have created a more acute shortage of energy. Delaying the measures that need to be taken in order to avoid a serious climate change, in particular a reduction of the emissions of carbon dioxide, will obviously make it more difficult to find ways and means towards a sustainable global energy supply system.

13.3.2 Carbon sequestration and storage

The Climate Convention repeatedly states that reducing emissions of greenhouse gases and enhancing their sinks are the means to stabilise the global climate. Only quite recently has the latter option received wide-spread attention. It is obvious that increasing the uptake of carbon dioxide by the oceans, the terrestrial systems, deep geological structures or aquifers might equally well serve the purpose of reducing the rate of increase of the atmospheric carbon dioxide concentration, but it is essential that long-term secure storage can be provided and the potential sink has to be large.

Enhanced *oceanic storage* is technically difficult and costly, and may also be illusive. Its efficiency depends on the ocean circulation which may change as a result of climate change. However, warmer surface waters will increase the vertical stability of the oceans and thereby possibly decrease the rate of carbon dioxide transfer back to the atmosphere, but nevertheless long-term secure storage cannot be guaranteed. This is not a viable possibility.

Terrestrial ecosystems including soils have in recent decades increasingly served as a significant natural sink for atmospheric carbon dioxide and thereby kept the air-borne fraction of the annual emissions at a rather low level, i.e. about 45%. The Kyoto Protocol recognises the possibility that increasing terrestrial storage is one way to fulfil commitments to reduce net emissions. It is, however, necessary to adopt a system that ensures that the efforts to enhance sinks do not bring about return flow to the atmosphere later, nor provide loopholes in the reporting system that would have to be established. It will be difficult to ascertain whether the reported uptake is assessed accurately because of the heterogeneity and temporal variability of the uptake of carbon dioxide by the terrestrial systems. It is also questionable whether uptake by terrestrial systems will provide adequate storage for a long time into the future. The stability of an enhanced terrestrial carbon reservoir cannot be guaranteed when the climate is changing (see Cramer *et al.* (2001)).

Geological storage in exhausted oil and gas fields as well as in aquifers below a depth of 800–1000 metres is an interesting possibility, but it has not yet been sufficiently investigated with regard to long-term stability and other possible environmental implications (see IPCC (2005)). The capacity of such storage sites might be very large, at least several hundred Gt C (probably more). The logistics required for the transfer of the carbon dioxide to the storage sites should not cause difficulties according to the IPCC analysis. The possibility is technically attractive for the energy industry, since the exploitation of deep storage fits well into its traditional structure and the technology being used. Rough estimates of the costs of this kind of mitigation indicate increases of up to about 100% for providing energy if coal is used as the primary energy source, i.e. of about the same magnitude as the price increase of crude oil that has been experienced in the last few years, to which an adjustment in society is under way. It is very clear that this possibility is quickly becoming a 'technology for the future' both in the USA and the EU.

Geological storage also offers another very interesting possibility in the context of using bioenergy. It can be exploited as a net sink at stationary installations, e.g. power stations producing electricity. The atmospheric carbon dioxide assimilated by photosynthesis would be brought into permanent storage in this way (see Azar *et al.* (2006)).

13.3.3 Other forms of geo-engineering

Crutzen (2006) has revived the earlier idea of using a reflective layer in the stratosphere to reduce solar radiation at the earth's surface. The basic idea is to inject sulphur dioxide into the stratosphere, which would quickly be transformed into sulphate particles by oxidation due to solar UV radiation. If injected at a level of 20–30 km, the particles would stay in the stratosphere for some years and a regular supply of sulphur dioxide into the stratosphere, up to a few million tons per year, might not require an impossible effort, even though a global distribution of this kind of a shield at the appropriate level might be difficult to achieve. The weakness of the proposal is, however, obvious. Once established and the carbon dioxide level thereafter allowed to increase, the shielding aerosol layer would have to be maintained permanently regardless of political developments on earth in order to prevent a rapid climate change occurring. Crutzen emphasised that this kind of measure should only serve as a final resort and should only be used if judged to be absolutely necessary. The political implications are, however, largely impossible to assess and the idea is therefore unrealistic.

In general, geo-engineering is not a viable solution because in most cases it is an illusion to assume that all possible secondary changes can be foreseen. Carbon storage in aquifers might, however, be such an exceptional possibility because the storage would be distributed among many separate aquifers and 'all the eggs would not be put into the same basket'. However, careful studies must precede the adoption of such measures.

13.3.4 Renewable energy

What then are the prospects for using renewable energy rather than fossil fuel energy as the world's prime source of energy? Renewable energy, above all *solar energy*, which is the source of all forms of renewable energy, will in the long term have to become the basic source of primary energy. How quickly can this be achieved?

The competitiveness of the energy market ensures that energy is produced as efficiently and cheaply as possible. Renewable energy technology has so far suffered in the tough competition with cheap fossil energy. About 80% of the primary energy used in the world today is therefore provided by fossil fuels (see Table 13.4), a figure that has been increased by the rapidly increasing demands from developing countries. The following brief overview highlights the present modest contributions from renewable energy resources to the total primary energy supplied and the reasons why (see UNDP (2004)).

Hydropower is a clean and cheap source of primary energy for the production of electricity that contributed about 2.3% of the total world energy used in 2001 and about 15% of the electricity provided in the world. The global resources are, however, limited and their contributions probably cannot be much more than tripled because of its inaccessibility of and the transfer losses from remote power plants. In comparison, the global demand for electricity is presently increasing by some 2% per year and will probably continue to increase for decades to come.

Table 13.4. *World primary energy supply, 2001 (as a percentage),*
see UNDP (2004)

Fossil fuels	79.4		
oil		35.1	
natural gas		21.7	
coal		22.6	
Nuclear energy	6.9		
Renewables	13.7		
large hydropower		2.3	
traditional biomass		9.3	
'new renewables'		2.1	
modern biomass			1.43
geothermal			0.51
small hydropower			0.09
wind electricity			0.04
solar photovoltaic			0.02
solar thermal			0.01
marine energy			<0.01

Salt energy is the utilisation of the chemical potential created between fresh water and salt water when rivers enter the sea. The osmotic pressure might be used for electricity generation and rough estimates show that the potential is not negligible. The key technical problem is the development of sturdy membranes allowing the potential to be transformed into useful energy. Estimates in Norway show that the potential is large enough to justify further technical development and it may be possible to use it to generate electricity, but salt energy will never become a major source.

Wind energy is still a small contributor to the global energy supply, but is increasing by 20–25% per year, primarily in Denmark, Germany and some states in the USA (e.g. California). Its possible future contributions should by no means be ignored, but it is still not economically competitive in most places and requires a rather open landscape to be publicly acceptable.

Photosynthesis today stores solar energy in the terrestrial systems in amounts that are about 8 times larger than the total primary energy used by humankind. About a quarter of this energy ends up in wood. Only 1–1½% of the total amount that is being stored in the terrestrial system is presently used as a source of energy by modern society, i.e. traditional biomass, wood, dung and biogas from waste dumps contribute some 9% of the total energy use. The emissions of carbon dioxide when burning biomass are certainly not different than when burning fossil fuels, except that we are returning to the atmosphere an amount of carbon dioxide that in the recent past was removed from the atmosphere by photosynthesis, i.e. the organic compounds that were formed are used as an energy source rather than letting this material decompose naturally.

The use of biomass can obviously be further enhanced a great deal. Its use is expanding in countries where forestry, timber and pulp production is an important industry, i.e. at boreal latitudes, and in the tropics. Biogas from decomposition of waste is preferably collected in densely populated regions in reasonably warm climates. Bioenergy will in any case be an important source of energy. In fact more than 20% of the total primary energy being used in Sweden stems from biomass, particularly waste from the forest and pulp industry.

Modern biomass, produced in plantations that are exclusively aimed at providing renewable energy, might become a considerably more significant energy source at northerly latitudes because of the extensive land areas available, i.e. in Canada, the Scandinavian countries and Russia, as well as in moist tropical countries where photosynthesis is rapid because of the warm and humid climate. Plantations of quickly growing species are supplying increasing amounts of energy, e.g. *Salix* at temperate and high latitudes for heating and power production, *sugarcane* and *corn* in tropical and subtropical countries for the production of ethanol which is used in the transport sector. In Brazil about twice as much biomass per unit area is obtained from sugarcane than from corn and the ratio of the energy in the final output to the energy used for its production is about 8 for sugarcane, and only about 1.5–1.8 for corn. It is further to be noted that the energy required for processing ethanol from sugarcane in Brazil is to a considerable degree supplied from bagasse, the waste product in ethanol production, while externally supplied energy, often from fossil fuels, is used to keep production going in the USA, where corn is the basic crop. In the USA, subsidies are required to make the process economically viable. Ethanol production based on corn is therefore hardly sustainable. The total production of ethanol is still only about 30 million tons annually, which represents a few per thousand of the global energy supply as given in Table 13.4. With sugarcane, the ethanol yield is up to about 6–8 t ha^{-1} in Brazil, which means that an area of about 50 000 km^2 is required to produce 30 million tons of ethanol from sugarcane plantations (Coelho *et al.*, 2005). It should be noted, however, that fertilisation will be required if the yields are to be maximised, which will lead to enhanced nitrous oxide emissions which will partly offset the reductions in the greenhouse effect (see Crutzen *et al.* (2007)). Also, other activities in the process of production and marketing of ethanol require energy and the net reductions of carbon dioxide emissions may be quite modest (see Patzek and Piementel (2005)).

Harvesting solar energy directly for heating and electricity generation and through the use of wind energy and other renewable energy sources stemming from solar radiation still plays a small role in the present global energy supply system ($<1\%$) and its contribution in absolute terms is increasing rather slowly. An important technical development in the exploitation of wind energy has markedly increased its potential and wind energy is becoming competitive with fossil fuels for electricity generation. In the future its role is likely to become increasingly competitive, even though its present contribution to the global energy needs is still only a few tenths of a per cent of the total. An expansion of solar electricity production will require substantial technical development and financial resources. Its contribution to the total energy supply is still next to negligible; research and technical development are required and are on the way. How rapidly will progress be made?

13.3.5 Nuclear energy

Nuclear energy will undoubtedly play a part in the world's future power supply system, not least because the development of renewable energy will take time. The approximately 440 reactors currently generate about 0.4 TW. Even though reactor effiency is being improved, at present they convert only about one third of the total energy produced into electricity; the rest is dissipated as waste heat. Nuclear power is currently supplying about 2.3% of the world's energy, about the same as hydropower. The waste heat is seldom used, even though there will certainly be incentives to harness more of it in the future. At present some 60 new reactors are being planned and in addition about another 110 have been proposed. There may thus possibly be some 150 new plants in operation in 2020 producing about 0.15 TW in addition to the present contributions of nuclear energy, i.e. an increase of about 40% in 15 years, which means 2–2½% per year. The demand for electricity is increasing at about that rate annually so if the nuclear industry expands as expected the percentage of the electricity provided by nuclear energy will remain the same over the next 15 years, i.e. at about 15%. It is obvious, however, that more rigorous international safety measures will be required if a major global expansion of nuclear energy is contemplated, especially since most of the supply of new energy will be needed in developing countries.

13.4 The supply of energy under the constraints of minimising climate change

13.4.1 The scope of the problem

The main conclusion of the analysis so far is thus simply as follows. The resources of fossil fuels in the world are large. An expanding world economy could to a considerable extent be based on a primary energy supply from fossil fuels well into the twenty-second century, even though a transition from conventional oil to natural gas and coal would be required and calls for substantial technical development and financial investments. But fossil fuels are non-renewable resources and must ultimately be replaced by other sources of primary energy. Increasing emissions of carbon dioxide might otherwise lead to a more than tripling of the preindustrial atmospheric carbon dioxide concentration. In addition to climate change other modifications of the global natural system might be initiated by a substantial increase in the atmospheric carbon dioxide concentration, e.g. acidification and warming of ocean surface waters with possible major destructive effects on marine ecosystems and a reduction in the role of the oceans as a sink for the increasing atmospheric carbon dioxide concentration.

The scope of the problem is most easily appreciated when it is realised that the world energy demand might double well before the middle of the century if the present pace of industrialisation continues, particularly due to the rapidly increasing demands from developing countries. Fossil fuels must not be the prime source of energy to meet this growing demand if more than doubling the carbon dioxide concentration is to be avoided. Their use would rather have to start decreasing within a few decades. A glance at Table 13.4 shows the huge gap that has to be filled through more efficient use of energy, the introduction of different forms of renewable energy, the capture and sequestration of carbon dioxide when using fossil fuels and possibly also an expansion of the use of nuclear energy. *The required measures would have to provide an energy supply twice the total world energy used today.* It is important in this context also to recall that the world population, presently about 6.4 billion, is expected to increase to about 9 billion in the next 50 years, i.e. the number of people added is expected to be more than the present populations of China and India combined.

The large population of the world has become a geophysical force that is transforming the pristine earth that was the cradle of the human race. According to Paul Crutzen, since about 1950 we have been entering a new geological epoch, we have been passing from the *Holocene* to the *Anthropocene*, i.e. the human-dominated geological epoch (Crutzen, 2002).

The development of a new global energy supply system involves much more than assessing and mapping the availability of new primary energy resources and the development of the technology required to exploit them efficiently. The more effectively we can make use of the energy that is available the better off we would obviously be. Because of the threat of a climate change, it is necessary to consider how the infrastructure of society might have to change, what the associated political implications might be and what measures should be taken. Technical development will have to be fast enough that the replacement of fossil fuels as the prime energy source is sufficiently rapid to keep the ongoing climate change within bounds. The problems confronting humankind will, however, surely not be resolved by a few magic international agreements. There is simply no 'silver bullet'. A number of parallel efforts must be pursued and it is important to ask the question: what can be done with already available technology, while the magnitude of the issue and its implications are becoming better understood and technical development accelerated (Pacala and Socolow, 2004)? In reality energy is presently not used efficiently, primarily because fossil energy has been very cheap. The increase in the price of energy during the last few years has stimulated its more efficient use and it seems likely that comparatively high prices for energy have come to stay. In that regard the market economy is pushing the development in the right direction. But the traditional infrastructure is a major obstacle in that it limits our ability to adjust quickly to new circumstances and people may also be affected very differently by the necessary transformations of society, a change that has already begun, although as yet only slowly.

13.4.2 Incentives aimed at reducing carbon dioxide emissions

For these very reasons economic incentives provide powerful mechanisms that might be employed more generally in order to reduce the present waste of energy. A trading scheme for emission permits is now being tested in the EU with the aim that it will be more generally introduced at the beginning of the first commitment period in 2008. But as yet it includes only emissions from stationary sources, such as power stations and the metallurgic industry. The basic idea is to stimulate a reduction of carbon dioxide emissions where it can be most easily achieved and at least cost. A limited number of tradable emission permits have been issued and distributed (either free of charge, or more appropriately through an auction). Enterprises that can reduce their need for fossil fuel energy by comparatively cheap means are able to sell some of their permits to those for whom a more efficient use of energy would be more expensive. In the long term a gradual decrease in the total number of permits would reduce the total use of energy produced by burning fossil fuels.

   This introduction of emission permits has led to a substantial increase in the price for electricity in Europe, which was expected because of the limited numbers of permits that were distributed. Tactics have also been developed that minimise the effects that are detrimental to the participants. Trade between EU countries and those not taking part is, however, becoming distorted to the disadvantage of EU countries. It is no surprise that there are protests against increasing energy prices, but these are just a manifestation of the ongoing increase in demand that in the long term will make the switch to non-carbon emitting primary energy relatively less costly than it would otherwise have been. An expansion of such a trading scheme to the world as a whole would be desirable, but the time is not yet ripe to do so because of the very different financial resources that characterise the international scene. It might actually be preferable to begin with the establishment of a set of regional markets in order to try out the basic idea further.

Another kind of effort to reduce emissions has emerged in the last few years. Companies have been formed that sell emission reduction certificates to private enterprises, organisations or even individuals. For the money received, they guarantee reductions of greenhouse gas emissions somewhere in the world, primarily in developing countries. The Norwegian government will, for example, enter into such an agreement to assist in reducing emissions by paying for such efforts. These kinds of initiatives are, of course, welcome but it remains unclear how effective they will be in the long term. In particular, governments must not use them as alternatives to reducing emissions 'at home'.

A few more specific examples will illustrate the role of market measures more clearly.

Increasing demand for more energy is most strongly felt in the transport sector. Energy efficiency in *transporting goods* on the road has improved substantially during recent decades. The reduction of fuel costs has been important in market competition, but the greater demand for transportation has anyhow increased total fuel consumption. Further reductions will primarily depend on whether fuel costs increase further or not; the open market has come to stay. The *private transportation* sector is also developing rapidly and is similarly increasing its share of the total energy use because of the increasing standards of living of most people in the world. There are more automobiles on the roads and aeroplanes in the air, and there is an urge amongst people in industrialised as well as in many developing countries to spend an increasing part of their spare time and a larger share of their increased earnings on travelling. On the other hand, it is likely that more efficient engines will significantly decrease the fuel consumption per unit distance travelled and slow the rate of increase of the energy use, but we do not know how quickly this might be achieved. *Hybrid cars* increase the efficiency of the fuel use by 30–40% and *electric cars* would do so further, but batteries with adequate storage capacity are not yet available, although technical development is on the way. The use of electric cars would obviously also reduce air pollution in densely populated areas, but it would imply an increasing demand for electricity to be produced without an increase of carbon dioxide emission. However, the rapidly expanding fleet of cars in the world means that we may not see a reduction in the global demand for fuel in the transport sector for quite some time, rather the opposite. This demonstrates the need for a change of life style, though this is not easily achieved. However, social tensions are increasing, particularly in the megacities of developing countries because of intolerable congestion that put limits of the future rate of growth but will most likely not stop it. The conflicts between traditional rural life and the hectic pace of industrialised societies are causing problems. The wishes of people in developing countries to follow in the footsteps of the inhabitants of industrialised countries means that the present number of cars in the world (almost one billion) might possibly have doubled by the middle of this century.

Another approach to reducing the use of fossil fuels in the transport sector would be to use biofuels more widely, especially ethanol. Plans are being developed in Brazil to increase its production capacity to supply about 5% of the fuel required for the expected fleet of cars in the world in 2025, which implies the production of at least 100 million tons of ethanol per year. Even such a major national effort would, however, lead to a reduction of the total amount of carbon dioxide emitted by the use of fossil fuels by just a little more than 1%. To supply the world fleet of cars towards the middle of the century with biofuels from the tropics would require a land area of about 3 million km^2, which corresponds to about 20% of all of the land in the world that is being used today for agriculture, a fact that again illustrates the scope of the problem that is confronting us.

The use of ethanol as a fuel in the transportation sector is being taken seriously in a number of other tropical countries. Ethanol is environmentally friendlier than gasoline and may well become a competitive source of energy in the world market, as gasoline prices rise. This might ease the potential conflicts between developed and developing countries in the negotiations of the structure of international treaties within the FCCC. Recall, however, the uncertainty about how much actually would be gained by replacing gasoline by ethanol in the transport sector, as pointed out in Section 13.3.4. The production of ethanol from corn, wheat, cellulose, etc. at middle latitudes is less efficient and hardly yet a rational solution for the future.

There are a number of means of increasing energy end-use efficiency in both industry and the private sector. The use of heat-pumps might well halve the requirement of energy for the provision of heat to public premises as well as private homes. Producing the electricity required without the emission of carbon dioxide remains and might well be a problem where renewable energy sources or nuclear energy are not available.

Better insulation wherever heating is required is an obvious measure on many occasions and will be economically more profitable the more energy prices increase. It is generally not adequately recognised that major savings can be achieved in this way. Considerable energy savings can also be achieved in the maintenance of good indoor air quality by using effective heat exchange arrangements for the ventilation systems.

These simple examples are just a few illustrations of the multitude of measures that can be used to reduce carbon dioxide emissions.

A hydrogen society is the name given to a society that uses hydrogen as a non-polluting fuel in the transport sector and more generally as an energy carrier from the place of production to final use. It should be recognised, however, that the use of hydrogen as the energy source for air traffic brings with it increased injections of water vapour into the stratosphere and enhances the formation of clouds in the stratosphere and upper troposphere, which might contribute to climate change. On the other hand, it would eliminate direct emissions of carbon dioxide wherever it is used, though it raises the basic question of which primary energy is to be used to produce the hydrogen (see National Research Council (2004b)). In the long run solar energy or other forms of renewable energy would be the preferred option, but this will not be possible for quite some time. Rather, the most plausible solution for the supply of the energy required for hydrogen production seems to be electricity from coal-based power plants combined with carbon sequestration and storage.

13.5 The need for a multidimensional approach

The specific examples described briefly above illustrate well the magnitude of the effort that is required to the stabilise atmospheric carbon dioxide concentration. The basic technology is largely available, and will most certainly be further developed and there will be more inventions in the decades to come, but finding a commonly agreed strategy and policy is more difficult. Excessive costs are feared and efforts are accordingly beginning only slowly, particularly so in many developing countries.

A useful attempt to be more specific about what might be needed during the next 50 years has been presented by Pacala and Socolow (2004). They set the goal of stabilising the atmospheric carbon dioxide concentration at 500 ppmv during the twenty-first century, which implicitly implies an equivalent carbon dioxide concentration of a little more than double the preindustrial one if radiative forcing due to other greenhouse gases is also taken into consideration. They assume that the stabilisation aimed for requires that the annual carbon dioxide emissions are maintained at about 7.0 Gt C until the middle of the century and then decline slowly during the remainder of the century. However, since their analysis was completed the emissions due to fossil fuel burning have increased markedly and the restrictions on emissions in order to reach the goal set by Pacala and Socolow would presumably be more stringent.

As a first step Pacala and Socolow define a number of action programmes, so-called 'wedges', that might be implemented during the next 50 years (see Figure 13.3). Each of these would achieve about 15% of the total reductions required to keep the annual carbon dioxide emissions constant at a level of about 7 Gt C for the next 50 years. Thus, seven wedges in total would be required to stabilise the emissions. This would imply a reduction of the business-as-usual carbon dioxide emissions by a total of 175 Gt C, i.e. seven wedges of 25 Gt C per wedge during a period of 50 years. Maintaining the atmospheric concentration at

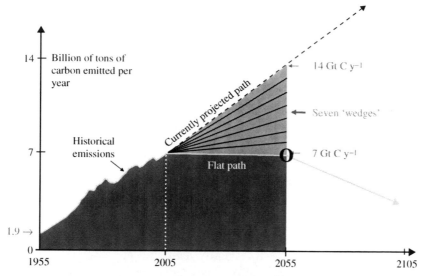

Figure 13.3   Assessment of measures required to achieve stabilisation of atmospheric carbon concentration at 500 ppmv. The upper curve shows 'business-as-usual' emissions according to Wigley *et al.* (2000). A constant annual emissions level of 7 Gt C for 50 years and thereafter a gradual decrease (as shown in the graph) would lead to the aimed-for stabilisation level in 2125. The 'gap' between the two curves is divided into seven 'wedges', each of which represents alternative sets of measures (as described in the text) that would contribute to the stabilisation (Pacala and Socolow, 2004).

about 500 ppmv thereafter would require a gradual decrease of emissions during the remainder of the century. The approach then requires the specification of a set of mitigation efforts, seven of which used in parallel would lead to stabilisation of the carbon dioxide concentration at about 500 ppmv in approximately the same manner as shown by the IPCC in 1995 and reproduced in Figure 8.1.

The authors describe some 15 wedges that illustrate different possibilities. Continued scientific research and technical development might make the requirements listed easier to achieve and also lead to the identification of other measures that might be undertaken. This is well illustrated by the following examples given by Pacala and Socolow:

- Improve the fuel economy of the approximately two billion cars expected in the world by 2050, from 30 to 60 miles per gallon (i.e. from 0.75 to 0.375 litre per 10 km).
- Reduce the annual distance travelled by each car from the present average of about 16000 to 8000 km per year.
- Reduce carbon emissions due to heating buildings and the electricity required for appliances by 25% in the next 50 years.

- Improve the efficiency of coal-fired power plants from 40 to 60% in the next 50 years (present average efficiency is about 32%).
- Introduce carbon capture and storage at 800 1-GW coal or 1600 1-GW natural gas power plants.
- Build 3500 carbon capture and storage installations of the kind being introduced at the Sleipner oil field in the North Sea.
- Add 700 GW nuclear power plants to replace present coal power plants, i.e. 14 reactors of 1000 MW should start operation each year, primarily in developing countries.
- Add about 2 million 1 MW wind mills (i.e. install 100 every day) which will ultimately occupy an area of about 300 000 km^2 on land and offshore.
- Increase ethanol production in Brazil to 100 times its current level, which would use 2.5 million km^2 of productive land.
- If photovoltaic electricity is technically available at reasonable costs, install 2000 GW peak photovoltaic facilities (i.e. about 500 times the present capacity) which would occupy about 20 000 km^2 of land that has good supply of sunshine.

In the mean time enhanced research and technical development will be needed to prepare for a decrease in emissions during the remainder of the twenty-first century.

Although the technology to achieve stabilisation of greenhouse gas concentrations is available, it is by no means clear what policy should be adopted to ensure that appropriate measures are undertaken. The use of macroeconomical models might be helpful in minimising the costs of stabilisation. A revision of the SRES scenarios is desirable in order to gain a better understanding of the stabilisation of greenhouse gas concentrations and to appreciate better the differences between developed and developing countries.

A similar approach can also be used to define national efforts to fulfil the commitments prescribed by a Climate Convention agreement on the reduction of national emissions with due regard to the special circumstances that individual countries will be facing. Other wedges may then also be explored. A particular complication arises. Many countries are not self-sufficient, while others are able to supply energy to the international market. The global character of the climate change issue means that global cooperation, negotiations and international treaties will still be required.

Already the first set of more elaborate macroeconomic models developed by the IPCC have shown that technical development in an open society, general awareness about the complex and difficult issues that lie ahead, and not least a stabilisation or possibly a gradual decrease in the world population, are key issues that must be addressed in the long-term perspective. An expansion of the IPCC macro-economic analyses has been completed by researchers at the International Institute for Applied Systems Analysis, IIASA, and some additional interesting conclusions have been drawn (Riahi *et al.*, 2006). Even their most optimistic scenario, which takes advantage of what has been learned from the previous IPCC analyses (scenario B1), reaches stabilisation of the enhanced greenhouse effect only at an equivalent carbon dioxide concentration of about 670 ppmv early in the twenty-second century, i.e. at a carbon dioxide concentration of about 600 ppmv, with the other human-induced greenhouse gases making up the rest. The authors comment that 'the B1 storyline describes a convergent world with a low global population growth that peaks in mid-century and declines thereafter [to 7.1 billion by 2100], but with rapid changes in economic structures towards a service and information economy, with reduction in material intensity and the introduction of clean and resource efficient technologies. The emphasis is on global solutions to economic, social and environmental sustainability, including improved equity.'

The basic scenario used results in a preference for measures related to

(a) energy conservation and efficiency improvement,
(b) fossil carbon sequestration and storage,
(c) nuclear energy,
(d) photovoltaic (solar energy),
(e) use of biomass, including carbon sequestration and storage,

because of their technical and economical advantages. An even more stringent policy is required if aiming at a lower stabilisation level. The scenario obviously supports the previous conclusions in this chapter that far-reaching and urgent action is required in order to keep the ongoing climate change within bounds. The authors further comment:

In this approach, while we certainly do not consider the B1 scenario 'likely' in view of current trends, we claim that it is perhaps the most likely scenario to yield low emissions and low vulnerability to climate change in a comprehensive assessment of uncertainties. Thus, even if challenged, we maintain the legitimacy of the 'convergence' theme that underlies the B1 scenario as a 'best case' scenario for climate policy assessment. We also maintain that the scenario, while being 'extreme' in the unfolding of existing trends, is not counterfactual (hence not implausible) with respect to historical experience, economic theory and the evidence put forward by the economic convergence literature once inherent data, measurement, and modelling uncertainties are taken into account.

We know little about the social consequences of the assumptions made in the course of developing this scenario, for example how a declining world population (particularly in developing countries) might come about and what it would imply, but such a decline is already happening in some EU countries (see Lutz *et al.* (2003)). This analysis and others of a similar kind must not be taken literally with regard to the more precise features of the outcome, but they should be seen as clear warnings about difficulties that lie ahead.

13.6 The economy of a transition to a sustainable energy supply system

The transition to a sustainable energy supply system for the world would certainly incur substantial costs, annually increased expenditure of the order of hundreds of billions US dollars for quite some time to come. This is seemingly a frightening sum of money, even for large industrial countries. This estimate was given in the third IPCC assessment (IPCC, 2001c). It was also the estimate given in a report from the Russian Academy of Sciences (Academia NAUK) to President Putin, when the Russian Federation was hesitating about ratifying the Kyoto Protocol and may have been the cause of the delay in the Russian Federation ratification.

These costs, as detailed in the IPCC assessments, may be stumbling blocks in negotiations, but the necessity for a major societal change in modern industrialised society must be viewed in a long-term and broad socio-economic context. It must be considered in relation to the regular renewal of the infrastructure of society and the annual increase of the gross domestic product of individual countries as well as the world as a whole (see Azar and Schneider (2002)).

Concern about a possible serious global climate change is obviously now becoming recognised at high political levels. Late in 2006 the UK government presented a detailed economic analysis of the climate change issue and its socio-economic impacts that had been requested by the UK prime minister (see Stern (2006)). The introductory sections of the report describe in some detail the expected changes of the global climate, largely based on the IPCC assessments, supplemented by references to some selected articles that have appeared in subsequent years. The picture painted is, however, rather exaggerated and not always supported by the IPCC assessments. The estimate of cost for future damages is thus not based on the more constrained IPCC estimates, as described in this and the previous chapter of this book.

Nevertheless, the Stern report is a forceful statement about the serious situation that humankind is facing, with the conclusion that it is urgent to act now. The report, however, rather underestimates the efforts of emission reductions required in order to avoid more than a 2 °C increase of the global mean temperature. It also ignores the fact that because of the inertia of the climate system overshooting the concentration requirements for stabilisation would lead to a warmer climate for many decades to a century and cause changes in the ecological systems that are at least partially irreversible. Nevertheless, the Stern report may actually be the most important sign that 2006 might be a turning point in addressing the global warming issue politically. The key question is: when and how will the necessary measures be enacted?

Development and renewal in the global society goes on constantly and is characterised by the successive build up, maintenance and regular change of its infrastructure, which is 'paid for' as part of the gross domestic product, even though most of that is being used to provide food, clothes, shelter and to satisfy other needs that are part of people's every day life. Now that the development of a long-term sustainable energy supply system is needed, the key questions are obviously: how quickly could this be achieved and how much of the increasing global productivity during coming decades will be required in order to accomplish such a change? In particular, what will it take to stabilise the concentrations of greenhouse gases in the atmosphere as part of a course towards a future sustainablility?

We do not know the answers to such questions (IPCC, 2007c), but early rough IPCC estimates yielded values for an annual expenditure of the order of 1–2% of the gross domestic product (see IPCC (2001)). These costs will not arise suddenly, but will occur gradually as worn-out plants are replaced. There is thus a need to allocate gradually increasing resources during a transition period of several decades in order to bring about such major changes.

However, the key parameter used for the long-term analysis of sustainable development should be the net domestic product (NDP), rather than the gross domestic product, which is commonly used today in reporting changes of the world economy. In other words, we must deduct from the gross domestic product the decrease in society's assets due to the ongoing exploitation of natural resources, many of which are finite, and the decline of the ecological services that are provided by land and water, which are often mismanaged.

The costs of a transition are by no means excessive if seen from a long-term perspective. It should be recalled that countries generally aim to increase their gross domestic product by about 2%, or preferably more annually, and have so far commonly succeeded in doing so. In a well-managed national economy even the net domestic product is likely to be positive. On the other hand, countries that maintain high living standards for their inhabitants by excessive exploitation of available natural resources are not on the track of sustainable development.

Azar and Schneider (2002) have pointed out that merely setting aside annually an additional fraction of a per cent of the annual production would in just a decade or two add up to the resources that are estimated to be required to build a sustainable energy supply system. However, they base their analysis on the comparatively large surplus of gross domestic product that is commonly reported. In many cases, however, the net domestic product might be negative, which implies that adequate investments for a sustainable future may lead to a necessary reduction in the resources available for consumption which might especially be the case in developing countries.

It should also be noted that the estimated costs have been arrived at assuming there is an open and free market. In light of the very large differences in economic power between the countries of the world the implications of this assumption must be carefully considered. A major change in the global infrastructure must be based on a vision of a more equitable world than exists today and an improved cooperative spirit that would be to the advantage of most, if not all, countries. But many poor developing countries might still suffer. It is therefore obvious that national sovereignty will still be an important factor in the intergovernmental negotiations that are required in order to settle on ways and means to reach agreements within the framework of the Climate Convention. Finally, an implicit and essential assumption is also that there will be no major disturbances in the development of a sustainable global society, i.e. that a steady and peaceful although modest increase in productivity can be maintained.

The transition to a new energy supply system is by no means a simple task. There is competition for the capital that is required to expand the infrastructure of society. A crucial matter is that the economic rewards of the increased productivity in industrialised countries do not necessarily end up in the hands of those that carry the responsibility for securing the future energy supply or sustainability in general. Rather, in a free market economy the earnings from increased productivity are primarily shared between the owners of profitable firms and their employees. The key political issue thus becomes one of facilitating the channelling of the capital required to support the efforts to secure a sustainable energy supply system in a manner that also reduces carbon dioxide emissions to the atmosphere. Such investments may well in the short term be less profitable than those having a more immediate public appeal. The importance of raising public awareness and the need for appropriate political interventions are obvious.

A major part of the Stern report is devoted to a discussion of economical means to develop and make use of available institutions in order to achieve a decarbonisation of the energy market. Some might consider raising taxes on energy in order to secure the funds required and give government the prime role in ascertaining what is a desirable development. This would mean that politicians made the decisions about the new energy supply system. This might prove a cumbersome and ineffective process. On the other hand, the introduction of tradable emission permits that has begun in the EU lets the market secure the effectiveness of the process. The price for energy would increase (as has already been the case) which would be a stimulus for improving, for example, the end-use efficiency of energy by consumers. Emission permits have initially been handed out free of charge to the users (the grandfathering principle). This might also have to be changed when the total number of emission permits is reduced gradually, which in turn would result in higher prices for the permits.

As yet, however, the European trading system only includes some energy-related activities and is so far operating only within the EU. It must be expanded further in order to work well. It is essential to do this in a manner that does not cause difficulties for poor countries in order to minimise international conflict. Developing countries will also have to pay the increased prices for energy, and it is likely that they will have to be given special allowances for some time to come. Another approach would be to create temporary regional markets that would later be brought together into a world market. Even though a free and open market for energy is essential it must be recognised that there are almost 200 independent countries in the world and that international negotiations will be required to determine the way emission permits are distributed and political distortions of the setup avoided.

The Stern report has brought the climate change issue to the attention of the highest political level in UK and accordingly it will be dealt with by the G8, presumably with some leading developing country representatives taking part in attempts to outline and agree a global strategy for dealing with the global energy issue, while simultaneously limiting a future change of climate. The political legitimacy of the scientific assessments that the IPCC has produced, and Stern has expanded upon and deepened in the field of economics, has been considerably strengthened. The IPCC AR4 (IPCC, 2007c) provides additional relevant information that is essential in this context.

The Stern report does not deal much with technology, nor what possible physical limitations there might be when trying to achieve sustainable development. There are still a number of open questions that need to be analysed more thoroughly, particularly how the north–south issue is going to be dealt with. The increasing world population will of course also lead to increasing energy demands. Biodiversity may be decreasing because of exploitation of the terrestrial biosphere.

13.7 Politics of securing a global sustainable energy supply system

The fact that the Kyoto Protocol came into force in 2005 and the issuing of the Stern report in 2006 brought the climate change issue into focus publicly during 2006. The twelfth conference of the parties to the Climate Convention was held in Nairobi in November 2006 and a more forward-looking attitude was hoped for by many. The Kyoto Protocol does not require developing countries to take on any quantitative commitments to reduce their greenhouse gas emissions before the first commitment period comes to an end in 2012. It is now important to adopt a long-term perspective that goes beyond a 10-year second commitment period, and focuses on 2050 as the year by which stabilisation of greenhouse gas concentrations should be achieved. Specificity of such aims was almost completely missing in the Kyoto agreement.

The previous discussion indicates that stabilisation of greenhouse gas concentrations at a level of 550 (equ) ppmv carbon dioxide concentration (thus including contributions of all greenhouse gases affected by human activities) is not possible, unless very substantial and immediate increases in energy end-use efficiency can be achieved globally in addition to early and very stringent constraints on greenhouse gas emissions.

Even putting the stabilisation goal at 600 (equ) ppmv will require early participation by all countries to keep the annual total emissions of carbon dioxide below a level of about 10-11 Gt C per year, and not to exceed global net per capita emissions of about 1.2-1.3 ton C towards the middle of the century.

Even though there might still be disagreements about a more precise formulation of the kind given above, the targets for a second commitment period will have to be considerably tougher than were agreed for the first one. As emphasised earlier in the chapter, the limits of the carbon dioxide emissions will have to be imposed as very specific restrictions in the plans for a new sustainable global energy supply system and economic instruments such as tradable emission permits should be brought more into focus in order to achieve the required emission reductions. It is also essential to assess as far as possible the future risks at local and regional scale that will be emerging as a result of expected changes of the large-scale features of global climate.

Political readiness to take another major step forward in combating climate change is now essential. Another decade of delays in reaching an agreement on the rather tough targets for a second commitment period would be unacceptable. It is no easy task to reach agreements on binding commitments that would be considerably tougher than the Kyoto commitments. Stakeholders in industrialised as well as developing countries may find themselves threatened by the changes that combating global climate change will bring. And a power struggle between the countries trying to secure their future energy resources will not promote long-term sustainable development.

The outcome of the twelfth conference of the parties to Climate Convention in Nairobi was rather disappointing. There was agreement on enhancing the support to developing countries in their efforts to protect themselves and to adjust to a forthcoming climate change, but the resources will be small compared with the likely needs. The EU was successful in getting support for a closer analysis of what has been achieved by the Kyoto Protocol, which will probably be revealing.

The preparations in 2006 for agreements on 'permissible' country emissions during a second commitment period were, however, not completed. The general notion was that extensive discussion will continue and would not lead to final agreements until 2008 or 2009. This means that implementation will not begin until late in the first commitment period. Further, it is essential that the USA becomes a partner in the development of this second phase towards a *global* climate regime. Might this really be possible before a new US President is elected in 2008?

Major developing countries, China and India in particular, are very aware of the fact that their emissions are rising rather quickly and that a stabilisation of the carbon dioxide concentration at 550 ppmv or below implies that global average emissions per capita might be 'permitted' to reach, but not exceed, 1.2–1.3 t C in an equitable world. Developed countries would also have to approach that same target, see Figure 13.2, which would mean an average reduction in their carbon dioxide emissions by about 60%, most of it before the middle of the century. It is very important that technical development be generously supported and that the stage is set for rapid industrial development. New technical development is essential for an early transition to a sustainable global energy supply system with modest impacts on the climate. How rapidly can new sources of primary energy become competitive alternatives to the still rapidly increasing use of fossil fuels? As emphasised repeatedly in this chapter the task ahead of us is a tough one.

Some concluding remarks

The importance of trustworthy scientific knowledge in climate negotiations

As a scientist, I have been engaged in the interplay between scientific analysis and politics for many years, and I have tried here to present an analysis of what has happened over the last 40 years and where we stand today.

The analysis has shown that a penetrating examination of the facts is an absolute necessity when trying to understand and deal with the major societal and political issues that confront us when we try to resolve the global climate change issue. The analyses must be accepted as trustworthy by the international scientific community and should therefore be carried out as far as possible as an independent and open scientific endeavour. The IPCC reports have been produced in a manner aimed at securing this status. They have largely been free from influence by politicians and stakeholders of different kinds. Nevertheless, government representatives have attended the final plenary sessions, when the summaries for policy makers were finally agreed and the extracts of the key scientific conclusions formulated in simple terms without compromising the basic scientific analyses in the supporting documents. This is a fundamental prerequisite for successful climate negotiations, and decisions about the joint efforts to be made by the parties to the Climate Convention. The development of a strict procedure of the kind that has been achieved by the IPCC has been essential and has also been taken as a model for other efforts of a similar kind. It is most important that this approach is also pursued in the future and is not distorted by increasing pressures from stakeholders and politicians in a world that is threatened by human-induced climate change and is still increasingly the subject of power struggles.

Nevertheless, initiatives for action have been few and on many occasions have not attracted sufficient support, partly for political reasons but also because the scope of the climate change issue has been grossly underestimated. As yet there are few signs that measures undertaken so far by the parties to the Climate Convention have had a significant global effect. The human-induced radiative forcing caused by greenhouse gas emissions has continued to increase essentially unaffected by international agreements so far.

It should be clear from this overview that there will be many difficulties on the road towards agreeing on joint action because of inadequate basic knowledge with regard to environmental consequences but may be even more so because of an inadequate appreciation of how society might be affected and respond, in both industrialised and developing countries. Building an awareness of the socio-economic implications of a global climate change remains a major responsibility for the scientific community.

The difficulty in analysing the climate change issue more closely is also due to the fact that climate change is interwoven with a number of other political issues that often differ from one country to another and whose importance is inadequately recognised.

The ongoing change of climate is partly hidden and largely irreversible and the changes that have been observed so far are only a part of what the greenhouse gases that have already been emitted might ultimately bring about. Slow and delayed action has markedly decreased the likelihood of being able to limit the enhanced equivalent carbon dioxide concentration to twice the preindustrial one.

The climate system is basically chaotic, but changes of its gross features as a result of human intervention can be foreseen with some confidence. Further research should focus on the determination of short- and medium-term risks that result from combinations of the different changes expected at the local and regional levels and the gross features of long-term changes.

The impacts of expected changes on people and societies need to be viewed in terms of the prevailing lack of equity and social justice in the world, in particular with regard to the differences between poor and developing countries on one hand, and rich and industrial countries on the other.

Developing countries will continue to aim at higher standards of living, even though this might lead to increasing requirements for energy that still for some limited time will have to be provided by increasing use of fossil fuels. Industrialised countries are likely to continue to strive to increase their productivity, and as far as we can judge now will adjust to a reduced use of fossil fuels at only a modest pace. The inertia of society remains the major difficulty that must be overcome.

Increasing trade and rapid globalisation will change the structure of global enterprises and the economic powers of the world are going to play an increasingly important role. How should the tasks ahead of us be formulated in terms of attractive future business opportunities in society in order to stimulate competition in efforts to mitigate climate change?

The challenge ahead of us is to develop a global energy supply system that will secure sustainable development for the countries of the world, and simultaneously to take the steps necessary in order to slow down and ultimately stop the on-going global climate change while it is still within reasonable bounds. It is essential to clarify better the constraints on global solutions for adequate future energy supply and support research and technical development to optimise our use of available natural resources efficiently.

Analysis shows very clearly that the necessary transition to a global energy supply system must be built around solar energy as the primary energy source. However, this path will probably not be competitive with the continued use of relatively cheap fossil fuels for still quite some time. This dilemma must be overcome and will require international agreements, government interventions and above all research and technical development. Carbon capture and storage will necessarily be one approach to be followed during a rather extended transition period in order to limit the increase in atmospheric greenhouse gas concentrations during the first half of the twenty-first century and to keep the global changes of climate within bounds while long-term solutions are being developed and put in place. Also, the increased price of energy has come to stay, and this will stimulate its more rational use. In any case,

A sustainable supply of renewable energy is the prime long-term goal.
A viable solution of the dilemma as described urgently requires joint action.
It cannot be achieved unless genuine cooperation is fostered in the world.

Notes

1. Nineteenth-century discoveries

1. Stockholm's Högskola was a private institution for higher education and research, founded by the city of Stockholm, and became later the University of Stockholm.
2. For a more detailed presentation of Svante Arrhenius and his role as a leader of the Stockholm Physics Society, see Rodhe *et al.* (1997), Crawford (1997).
3. See Crawford (1996).
4. Excerpts from a lecture given at Stockholm Högskola, 3 February 1896 (Arrhenius, 1896b).

2. The natural carbon cycle and life on earth

1. For a more detailed account see Schneider-Carius (1955).
2. A note by Ausubel (1983) provides some interesting quotes by Lotka (1924) and Kostitzin (1935).
3. See also From and Keeling (1986).
4. $1 \, Gt = 10^9 \, tons = 10^{15} \, g = 1 \, Pg$.
5. See IPCC (2000a), Chapter 1.
6. The presentation has been based on the data provided by: Heimann (1997); IPCC (2000a), Chapter 1; IPCC (2001a), Chapter 3; and Field and Raupach (2004). Small differences between the syntheses by different authors are due to the uncertainties of observations and the means of deducing the numbers that appear in the figure. It should further be noticed that the sizes of reservoirs and fluxes in terms of carbon dioxide rather than carbon are obtained by multiplication by 3.67.
7. See IPCC (2001a), Chapter 3, Field and Raupach (2004).
8. See further IPCC, (2001a), page 39.
9. See Nilsson *et al.* (2003).

3. Global research initiatives in meteorology and climatology

1. President Kennedy's appeal to the UN resulted in the adoption of Resolution, 1721 (XVI), December 1961, which recommended: '. . . members and WMO to study measures to advance the state of atmospheric sciences and technology in order to improve existing weather forecasting capabilities and to further the study of the basic physical processes that affect climate.'
2. The resolution asked WMO '. . . to develop in greater detail its plan for an expanded programme to strengthen methodological services and research . . .' and invited ICSU

through its unions and national academies '. . . to develop an expanded programme of atmospheric science research which will complement the programme fostered by the World Meteorological Organisation.'

3. A more detailed account of the international planning work that ultimately led to the establishment of GARP has been given in Bolin (1986).

4. See National Academy of Sciences (1966).

5. The conference took place at Skepparholmen, near Stockholm. The proceedings were published as: ICSU/IUGG – CAS – COSPAR (1967).

6. An agreement between ICSU and WMO on the formation of GARP was reached in Rome in 1967. The task given to the JOC is reproduced in the report of the first session of JOC, April 1968.

   The 12 members of the JOC elected in 1967 were: Professor B. Bolin, University of Stockholm, Sweden, (*chairman*); Academician V. A. Bugaev, Meteorological and Hydrological Service, Moscow, USSR; Professor F. Möller, University of Munich, West Germany; Professor A. S. Monin, Institute of Oceanology, Moscow, USSR; Professor P. Morel, University of Paris, France; Professor Y. Ogura, University of Tokyo, Japan; Dr P. R. Pisharoty, Meteorological Service, Poona, India; Dr C. H. B. Priestley, CSIRO, Melbourne, Australia; Mr J. S. Sawyer, Meteorological Office, Head of Research, London, UK; Professor J. Smagorinsky, US Weather Service, Washington DC, USA; Professor R. W. Stewart, University of British Columbia, Canada; Professor V. E. Suomi, University of Michigan, Wisconsin, USA; Professor R. V. Garcia, formerly professor of University of Buenos Aires, who served as secretary of the Committee.

7. The task is reproduced in full in JOC (1968).

8. The results of the FGGE were presented at a study conference held in Geneva in 1985 (JOC, 1986).

9. The conference took place in Stockholm between 28 June and 16 July 1971 and the results were published as SMIC (1971). Stephan Schneider served as the secretary of the conference.

10. As of 1972 the composition of the JOC changed in that Professor F. Möller, Professor A. S. Monin, Professor Y. Ogura and Dr C. H. B. Priestley had been replaced by Professor K. Gambo, University of Tokyo, Japan, Professor K. Hasselmann, Max Planck Institute for Meteorology, Hamburg, Germany, Academician A. M. Oboukhov, Institute for Atmospheric Physics, Academia NAUK, Moscow, USSR, and Dr G. B. Tucker, CSIRO, Melbourne, Australia.

11. The conference was held in Washington DC in April 1978. It was organised jointly by JOC and the US committee for GARP. The proceedings were published as JOC (1979). The members of JOC had also changed in the meantime. From 1976 the members were: Dr P. K. Das, India, Professor K. Gambo, Japan; Dr J. T. Houghton, UK, Dr C. Leith, USA, Professor P. Morel, France, Academician A. M. Oboukhov, USSR, Professor M. Petrossiants, USSR; Professor J. Smagorinsky, USA (*chairman*), Professor R. W. Stewart, Canada, Dr G. B. Tucker, Australia, (*vice-chairman*), Dr A. Wiin-Nielsen, European Centre for Medium Range Forecasting, UK; Professor J. Woods, Federal Republic of Germany.

4. Early international assessments of climate change

1. See National Academy of Science (1975). A more detailed account of early analyses in the USA has been given by Hecht and Tirpak (1994).

2. See Swedish Government proposition 1975/76: No. 30 to the Swedish Parliament on energy policy.

3. National Academy of Sciences (1979). The nine members of the assessment group were: J. Charney, Massachusetts Institute of Technology (*chairman*), A. Arakawa, University of California, B. Bolin, University of Stockholm, R. Dickinson, National Center for Atmospheric Research, R. Goody, Harvard University, C. Leith, National Center for Atmospheric Research, H. Stommel, Woods Hole Oceanographic Institution, C. Wunsch, Massachusetts Institute of Technology.

4. National Academy of Sciences (1982, 1983). In the latter report the impacts of climate change were exemplified; e.g. P. E. Waggoner, 'Agriculture and a climate changed by more

carbon dioxide'; R. R. Revelle and P. E. Waggoner, 'Effects of a carbon-dioxide-induced climatic change on water supplies in the Western United States'; R. R. Revelle, 'Probable future changes of sea level resulting from increased atmospheric carbon dioxide'; and T. C. Schelling, 'Implications for welfare and policy'.
 5. See Bolin *et al.* (1986).
 6. The manuscript of an article by Ramanathan and colleagues (later published as Ramanathan *et al.* (1985)) was made available to the assessment team in 1984.
 7. The conference was organised in Villach, Austria, in October 1985 by the Austrian Government.
 8. See Boehmer-Christiansen (1994a,b).
 9. The conference statement can be found at the beginning of the assessment report, Bolin *et al.* (1986).
10. The group consisted of K Hare, Canada (*chairman*), B. Bolin, Sweden, G. Golitsin, USSR, G. Goodman, Sweden and UK, M. Kassas, Egypt, S. Manabe, USA, Gilbert White, USA. Dr Goodman was the Director of the Beijer Institute on Human Ecology at the Royal Academy of Sciences in Sweden.

5. Setting the stage

 1. Quotes are from the report by the WCED (1987).
 2. For more details see Benedick (1991).
 3. For more details see Hecht and Tirpak (1994) and Agrawala (1998).
 4. See Hansen (1988).
 5. See IPCC (1988). The following countries attended the session: *Europe* (11): Denmark, Finland, France, Italy, Malta, The Netherlands, Norway, Sweden, Switzerland, West Germany, UK; *Eastern Europe* (1): USSR; *North America* (2): Canada, USA; *Latin America* (2): Brazil, Mexico; *Middle East* (2): Israel, Saudi Arabia; *Africa* (5): Algeria, Kenya, Nigeria, Senegal, Tanzania; *Asia* (2): China, India; *Pacific* (3): Australia, Japan, New Zealand.
 6. See further Hecht and Tirpac (1994).
 7. The UN General Assembly Resolution 43/53, 1988: the UN General Assembly requested 'the Secretary-General of WMO and the Executive Director of UNEP together with the IPCC immediately to initiate action leading, possibly within the next eighteen months, to a comprehensive review and recommendations with respect to

   (a) the state of knowledge of the science of climate and climate change, with special emphasis on global warming,
   (b) programmes and studies of the social and economic impact of climate change particularly global warming,
   (c) possible policy responses by Governments and others to delay, limit or mitigate the impact of climate change,
   (d) relevant treaties and other legal instruments dealing with climate,
   (e) elements for possible inclusion in a future international convention on climate.'

 8. The creation of the IPCC and its role for the later establishment of the Framework Convention has also been described by Hecht and Tirpak (1994).

6. The scientific basis for a climate convention

 1. See TATA Energy Research Institute (1989).
 2. See IPCC (1989b).
 3. Presentation by Prime Minister Margaret Thatcher at the Royal Society, 27 September 1988, Prime Minister's Office, 10 Downing Street, London.
 4. See the Noordwijk Declaration, 1989, published by the Dutch Government, also reproduced in the report of the third session of IPCC (IPCC, 1990d).
 5. See IPCC (1990d).
 6. See Bromley (1990). Dr Bromley was scientific advisor to President Gorge Bush.

7. See IPCC (1990a).
8. The key paper on the role of aerosols for the heat balance of the atmosphere did not appear until 1991; see Charlson *et al.* (1991).
9. See IPCC (1990b).
10. See Budyko and Izrael (1987).
11. Letter from J. Houghton to Yuri Izrael, 23 March 1989 (private communication).
12. A more detailed analysis of the research on climate change in the USSR and the methodology of Budyko was carried out in a joint US–Soviet research project (MacCracken *et al.*, 1990).
13. See IPCC (1990c).
14. See IPCC (1990e, 1992a). The desirability of producing a synthesis report was recognised early in the IPCC discussions.
15. More detailed accounts of the negotiations that led to the creation of the Climate Convention have been brought together by Mintzer and Leonard (1994).
16. UN General Assembly Resolution 45/212, 1990.
17. See IPCC (1991a).
18. *Introduction*

  1. The IPCC shall concentrate its activities on the tasks allotted to it by the relevant WMO Executive Council and UNEP Governing Council resolutions and decisions.

  Organisation

  2. The IPCC Bureau shall reflect balanced geographic representation. IPCC Working Groups and any task forces established by Plenary shall reflect balanced geographic representation with due consideration for scientific and technical requirements.
  3. IPCC Working Groups and any task forces constituted by the IPCC Plenary shall have clearly defined and approved mandates and work plans as established by the Plenary, and shall be open ended.

  Participation

  4. Invitations to IPCC Plenary, Working Groups and task force sessions shall be extended to Governments and other bodies by the Chairman of IPCC.
  5. Experts from WMO/UNEP Member countries or international, intergovernmental or non-governmental organisations may be invited in their own right to contribute to the work of the IPCC Working Groups and task forces. Governments should be informed in advance of invitations extended to experts from their countries and if they wish may nominate additional experts.

  Procedures

  6. In taking decisions, drawing conclusions, and adopting reports, the IPCC Plenary and Working Groups shall use their all best endeavours to reach consensus. If consensus is judged by the relevant body not possible: (a) for decisions on procedural issues, these shall be decided according to the General Regulations of the WMO; (b) for conclusions and adoption of reports, differing views shall be explained and, upon request, recorded.
  7. Conclusions drawn by the IPCC Working Groups or task forces are not official IPCC views until they have been discussed and accepted by the IPCC Plenary.
  8. Invitations to IPCC Plenary, Working Groups and task force sessions should be extended at least six weeks in advance.
  9. Major reports, basic documents and other available reports for consideration at IPCC Plenary and Working Group sessions shall be made available by the IPCC Secretariat four weeks in advance, to the extent possible in all official UN languages. Working papers shall be circulated as far in advance as possible.
  10. Interpretation into all official UN languages should be provided for all IPCC Plenary and Working Group sessions.
  11. The sessions of IPCC Working Groups and task forces shall be co-ordinated with other international meetings, including sessions of the INC and UNCED Preparatory Committee.
  12. These principles are to be reviewed at least annually and amended as appropriate.

19. Letter of 31 January 1991 to members of the IPCC Bureau.
20. See IPCC (1992b).
21. See IPCC (1992c).
22. See IPCC (1991b) and Chapter 7, note 1.
23. International Gas Union Congress 1991, see Bolin (1991).
24. World Petroleum Congress 1991, see Bolin (1992).
25. His conclusions were also partly based on the work by Lindzen (1990).
26. The underlying analysis (see Pepper *et al.* (1992)) had been completed very late and had not been available to all participants.
27. For the details of the Convention text reference is made to The United Nations (1992).
28. Annex-I countries are listed in an annex to the Convention and comprise all OECD countries and in addition countries in economic transition, i.e. eastern European countries and former republics of the USSR.

7. Serving the Intergovernmental Negotiating Committee

1. Speakers at this meeting were: P. Gosh, TERI, India: Models, policy instruments and equity perspectives from economics; J. C. Hourcade, CNRS, France: L'analyse économique des outils pour cerner les implications des stratégies anti-effet de serre; M. Grubb, Royal Institute for International Affairs, UK: Uses and limits of economic analyses in climate change; R. Richels, Electric Power Research Institute, USA: Possible uses of macroeconomic models in the greenhouse debate.
2. The following Bureau members were elected:

IPCC	Vice-chairmen:	Professor Y. Izrael, Russian Federation
		Dr Al-Gain, Saudi Arabia
Working Group I	Cochairmen	Sir John Houghton, United Kingdom
		Dr. L. G. Meira, Brazil
	Vice-chairmen	Professor Ding Yihui, China
		Dr H. Grassl, Germany
		Mr M. Seck, Senegal
Working Group II	Cochairmen	Mr Robert Reinstein, USA
		(later replaced by Dr R. Watson, USA)
		Dr M. Zinyowera, Zimbabwe
	Vice-chairmen, also Cochairmen of the subgroups	
	Subgroup A	Dr M. Parabrahmam, India,
		Dr K. Yokobori, Japan
	Subgroup B	Dr P. Vellinga, The Netherlands
		Ms M. Martha Perdomo, Venezuela
	Subgroup C	Dr O. Canziani, Argentina
		Dr M. Beniston, Switzerland
	Subgroup D	Dr M. Petit, France
		Mr A. Hentati, Tunisia
Working Group III	Cochairmen	Ms E. Dowdeswell, Canada
		(later replaced by Dr J. Bruce, Canada)
		Dr H. Lee, Republic of Korea
	Vice-chairmen	Professor R. Odingo, Kenya
		Mr T. Hanisch, Norway
Regional representatives:	Africa	Dr J. Adejokun, Nigeria
	N & C America	Dr F. Moros, Cuba
	Asia	Dr H. Nasrallah, Kuwait
	SW Pacific	Dr M. Tegart, Australia
	South America	Dr N. Sabogal, Colombia
	Europe	Dr B. Perez, Spain

3. Executive Director of the Council for Effective Climate Change Policies (a non-governmental organisation with an obvious political agenda).
4. A more detailed analysis of the interplay between the scientific community and representatives of governments at the time is given in Bolin (1994a) and the continued work was outlined in Bolin (1994c).
5. See *The New Scientist*, 11 June 1993.
6. See also Section 8.3.
7. Articles 3.1 and 4.7 of the FCCC.
8. See Nordhaus and Yohe (1983), Nordhaus (1991) and Cline (1992).
9. The Huygens Lecture presented in The Hague, 16 November 1993: see Bolin (1993b).
10. The letter was signed by the executive directors of the Global Climate Coalition and the Climate Council, John Shlaes and Don Pearlman respectively, and the presidents or vice-presidents of the American Association of Railroads, American Petroleum Institute, National Coal Association, National Association of Manufacturers, American Automobile Manufacturers Association and Edison Electric Institute. It was sent to Mr R. Pomerance, US State Department, and Dr R. Watson, White House Office of Science and Technology, who were US delegates to the working group session. Copies of the letter were also sent to the Secretaries of the US Departments of Treasury, Agriculture, Commerce, Energy, to the Environmental Protection Agency, a number of key individuals in the government departments and agencies, and some selected members of the US Senate and House of Representatives.
11. Dr H. Linden, Illinois Institute of Technology, executive advisor, Gas Research Institute, Professor W. A. Nierenberg, former director of Scripps Institute of Oceanography, Dr F. Seitz, former president of the US National Academy of Sciences, Professor S. F. Singer, University of Virginia, and Dr C. Starr, founding president, Electric Power Research Institute.

8. The IPCC Second Assessment Report

1. See United Nations (1992).
2. I met with representatives from the Global Climate Coalition: Barbara Rosewiez, *Wall Street Journal*; Pam Zurer, *Chemical and Engineering News*; Nick Sundt, *Energy Economics and Climate Change*; Gary Lee, Curt Suplee and Sussan Okie, *Washington Post*; Greg Easterbrook, *Newsweek*.
   At the US Congress I met with Karen McCarthy from the House of Representatives, while a meeting with the Energy Committee of the Senate was cancelled. On the other hand, five representatives from the insurance industry were anxious to have a discussion and came with Frank Nutter in the lead.
3. See the editorial (by John Maddox) in *Nature*, 16 March, 1995. 199–200. A response by the chairman of Working Group I and myself was published a few weeks later.
4. The new information referred to was available in draft form and had been submitted to *Nature* for publication. See Santer *et al.* (1996).
5. Santer and Wigley, the two lead authors of the IPCC chapter, were key authors of the *Nature* article and they were undoubtedly in the vanguard of this kind of analysis at the time.
6. As a curiosity, it might be interesting to note that there is a major error in Figure 2 of the Working Group II summary for policy makers in that the two ecosystems 'Savannah, dry forests, woodland' and 'Tropical forests' have been interchanged, but I have not seen this corrected anywhere in the IPCC publications.
7. Convening lead authors: H Ishitani, Japan ; T. B. Johansson, Sweden.
   Lead authors: S. Al-Khouli, Saudi Arabia; H Audus, IEA; E. Bertel, IAEA; E. Beavo, Venezuela; J. A. Edmonds, USA; S, Frandsen, Denmark; D. Hall, UK; K. Heinloth, Germany; M. Jefferson, WEC; P. de Laquil III, USA; J. R. Moreira, Brazil; N Nakicenovic, IIASA; Y. Ogawa, Japan; R. Pachauri, India; A Riedacker, France; H.-H. Rogner, Canada; K. Saviharju, Finland; B Soerensen, Denmark; G. Stevens, OECD/NEA; W. C. Turkenburg, The Netherlands; R. H. Williams, USA; Zhou Fengqi, China.

8. World Commission on Environment and Development (1987). A related (somewhat stronger) concept is that 'each generation is entitled to inherit a planet and a cultural resource base at least as good as that of previous generations'.

9. *Top-down models* are aggregate models of the entire macro-economy that draw on analyses of historical trends and relationships to predict the large-scale interactions between the sectors of the economy, especially the interactions between the energy sector and the rest of the economy. They typically incorporate relatively little detail on energy consumption and technological change, compared with bottom-up models.
   Bottom-up models incorporate detailed studies of the engineering costs of a wide range of available and forecast technologies, and describe energy consumption in great detail. However, compared with top-down models, they typically incorporate relatively little detail on non-energy consumer behaviour and interactions with other sectors of the economy.

10. Enting *et al.* (1994); IPCC (1995a), Chapter 1; IPCC (1996a).

11. Members of the writing team: chairman of the IPCC, Bert Bolin; the cochairmen of the three working groups; and the heads of the three technical support units; R. A. Moreno, Cuba, lead author, Working Group III, energy supply, mitigation options; T. Banuri, Pakistan, lead author, Working Group III, equity and social issues; Z. Dadi, China, lead author Working Group III, emission scenarios; B. Gardener, USA, Global Climate Coalition; industrial expertise; J. Goldenberg, Brazil, lead author, Working Group III, economy; J.-C. Hourcade, France, lead author, Working Group II, costs and mitigation; M. Jefferson, UK, lead author Working Group II, energy supply, mitigation; J. Melillo, USA, lead author, Working Groups I and II, ecology; I. Minzer, USA, lead author, Working Group III; R. Odingo, Kenya, vice-cochair, Working Group III; M. Parry, UK, lead author, Working Group II, impacts and adaptation; M. Perdomo, Venezuela, vice-cochair, Working Group II, health and water resources; C. Quennet-Thielen, Germany, chief IPCC delegate; J. Stiglitz, USA, member of the Council of Economic Advisors to the US President; P. Vellinga, The Netherlands, vice-cochair, Working Group II, coastal zones; N. Sundararaman, secretary of the IPCC.

12. Letter, August 16, 1995, *from* the Global Climate Coalition (J. B. Schlaes and B. Gardiner), The Climate Council (D. H. Pearlman), the American Automobile Manufacturers Association (R. H. McFadden), the Edison Electric Institute (R. A. Beck), and the National Mining Association (C. D. Holmes), *to* the US State Department (T. Wirth, R. Pomerance), the US Department of Energy (H .R. O'Leary), and the US Environmental Protection Agency (D. M. Gardiner, D. A. Tirpak).

13. See IPCC (1996d), which is a very readable and easily comprehended report that captures the state of knowledge in the latter part of 1995 in the form of an integrated treatment.

9. In the aftermath of the IPCC second assessment

1. See Spencer and Christy (1993), Christy *et al.* (1995), IPCC (1996a).

2. See Hurrell and Trenberth (1997), Wentz and Schabel (1998).

3. See Western Fuels Association, INC (1996) *State of the Climate Report.* The report was supported financially by Western Fuels Association (which primarily supplied coal to the western parts of the USA) and was edited by Patrick J. Michaels, professor of environmental sciences at University of Virginia.

4. Reproduced in the *Bulletin of the American Meteorological Society,* see American Meteorological Society (1996). Dr Frederick Seitz was professor emeritus at the Rockefeller University, New York and former president of the US NAS.

5. See Santer *et al.* (1996). The IPCC procedure used on the occasion is described in Chapter 8. Compare also with the account of the Madrid session.

6. The letter was signed by Dr Susan Avery, chairwoman and Dr Richard Anthes, president of the University Corporation for Atmospheric Research, Boulder, CO, and Dr Paul Try,

president and Dr Richard Hallgren, executive director of the American Meteorological Society; see American Meteorooogical Society (1996).

7. In addition to Frederick Seitz, the George Marshall Institute leadership consisted of H. H. Linden, Illinois Institute of Technology, W. A. Nierenburg, director emeritus, Scripps Institute of Oceanography, S. F. Singer, professor emeritus of environmental sciences, University of Virginia, and C. Starr, funding president of Electric Power Research Institute. (It is of interest to note that Professor Singer was closely associated with Professor Patrick Michaels at University of Virginia.) Cosignatories of the letter were the Senators F. H. Murkowski, L. E. Craig, L. Faircloth, S. Abraham, L. Pressler, C. Burns, J. Helms, D. Nickles, J. B. Johnston, W. H. Ford, J. B. Breaux, R. C. Byrd, H. Heflin and B. L Dorgan.

8. Statement by the US Principal Delegate to the Second Conference of the Parties to the FCCC, T. Wirth. The US policy on global climate change as pursued by the US Administration was later presented by Tim Wirth to the subcommittee on international economic policy, export and trade promotion of the Senate Committee on Foreign Relations, 19 June, 1997.

9. See United States 104th Congress, Committee on Science (1996).

10. Dr J. Mahlman was director of the Geophysical Fluids Dynamic Laboratory, Princeton University, NJ.

11. See Hasselmann (1997a,b).

12. IPCC (1996f); IPCC (1997a); IPCC (1997b); IPCC (1997c).

13. See report to the fourth session of SBSTA (IPCC, 1996g); report to the fifth session of SBSTA (IPCC, 1997d).

14. See also the section on stabilisation of greenhouse gas concentrations, Chapter 8.3.

15. Letter from Mr Zou Jingmeng, principle delegate of China to IPCC, dated 3 June 1997; private communication.

16. Letter to Mr Zou Jingmeng, 26 October 1997; private communication.

17. Remarks by the US President on Global Climate Change at the National Geographic Society, Washington, DC, Office of the Press Secretary, The White House, 22 October, 1997.

18. B. P. Flannery, senior advisor at Exxon Research & Engineering Co, Global Climate Change presentation at Ghent, Belgium, 22 November, 1997 (available from the Exxon Research and Engineering Corporation).

19. Members of the nominating committee were the executive director of UNEP, the assistant secretary general of the WMO, and one member from each of the six regions as defined by WMO: Africa, Asia, Latin America, North America and the Caribbean, The Pacific Region, and Europe.

20. The composition of the Bureau for the next four years was:

	Chairman:	R. Watson
	Vice-chairmen:	K. Seiki (Japan), R. Odingo (Kenya), R. Pachauri (India), G. Meira Filho (Brazil), Y. Izrael (Russian Federation).
Working Group I	Cochairmen:	Sir John Houghton (UK), Ding Yihui (China).
	Vice-chairmen:	Baruhani (Tanzania), H., Nasarallah (Kuwait), A. Ramirez (Venezuela), J. Stone (Canada), J. Zillman (Australia), F. Joos (Switzerland).
Working Group II	Cochairmen:	O. Canziani (Argentina), J. McCarthy (USA).
	Vice-chairmen	A. Ndiaye (Senegal), A. Majeed (Maldives), S. B. Abdallah (Tunisia), J. Pretel (Czeck Republic), M. Manning (New Zealand), M. Petit (France).
Working Group III	Cochairmen	B. Metz (The Netherlands), O. Davidson (Sierra Leone).
	Vice-chairmen	E. Jochem (Germany), M. Munassinghe (Sri Lanka), E. Calvo (Peru), R. P. Madruga (Cuba), R. T. M. Sutamihardia (Indonesia), L. Lorentsen (Norway).

10. The Kyoto Protocol is agreed and the third assessment begun

1. International Energy Agency (1997).
2. Petition Project, PO Box 1925, La Jolla CA, supported by Oregon Institute of Science and Medicine, Oregon, 97523 (Arthur B. Robinson, Sallie L. Balliunas, Willy Son and Zachary W. Robinson) and George C. Marshall Institute, Washington.
3. The following note was published in Washington Post, 1 May, 1998: 'Former National Academy of Sciences president Frederick Seitz sent an eight-page paper in March to thousands of scientists, asking them to sign a petition against the Kyoto protocol on global warming saying the science supporting global warming 'is still uncertain', Senator Chuck Hagel (R-Neb.) told the House hearing last week of the extraordinary response to Seitz's letter. 'Three days ago, 15 000 scientists released a petition urging the rejection of the Kyoto Protocol,' Hagel said. 'Nearly all of these 15 000 scientists have technical training suitable for evaluating climate research data' – There was one name, though, that caught Ohio Democratic Rep. Dennis J. Kucinich's eye. He noted that the petition was signed by a Dr Jeri Halliwel. 'Perhaps she is a scientist,' Kucinich said, 'but she is also much better known as Ginger Spice of the Spice Girls.' Kucinich said later he was still checking for Posh, Baby and Scary and was heartened to see their involvement.'
4. *Wall Street Journal*, 16 April, 1998. See also *Business Week* magazine, 16 December, 2005: 'It's time to go on a low-carbon diet'.
5. *The Economist*, 20 December 1997, Environmental scares.
6. See http:\\www.john-daly.com artifact.htm.
7. IGBP Terrestrial Carbon Working Group (1998).

11. A decade of hesitance and slow progress

1. Bolin *et al.* (2000).
2. The IPCC has used the following words where appropriate to indicate judgmental estimates of confidence: *virtually certain* (greater that 99% chance that the result is true); *very likely* (90–99% chance); *likely* (66–90% chance); *medium likelihood* (33–66% chance); *unlikely* (10–33% chance); *very unlikely* (1–10% chance); *exceptionally unlikely* (less than 1% chance).
3. See Mann *et al.* (1995, 1999), Crowley (2000).
4. See McIntyre and McKitrick (2003).
5. See National Research Council (2004a) and IPCC (2007a).
6. See also Fu and Johansson (2004) and Fu *et al.* (2006).
7. See Pollach *et al.* (1998).
8. A thermo-haline circulation is driven by density differences of sea water due to temperature and salinity gradients usually created by ocean atmosphere interactions, i.e. heat exchange, evaporation and rainfall.
9. 1 EJ = 10^{18} J $\sim$ 250 TW h. The total annual use of commercially produced energy in the world is presently about 500 EJ $\sim$ 125 000 TW h. See also Table 13.3. Burning 1 Gt C produces about 50 EJ of energy.
10. Reserves (of fossil fuels) are those occurrences that are identified and measured as economically and technically recoverable with current technologies and prices. Resources are those estimated occurrences, not yet discovered and identified and thus also with less certain geological and/or economic characteristics.
11. The other members of the committee were: E. J. Barron, Pennsylvania State University, University Park, R. E. Dickinson, Georgia Institute of Technology, Atlanta, I. Y. Fung, University of California, Berkeley, J. E. Hansen, NASA/Goddard Institute for Space Studies, New York, T. R. Karl, National Climate DATA Center, Asheville, R. S. Lindzen, Massachusetts Institute of Technology, Cambridge, J. C. McWilliams, University of California, Los Angeles, F. S. Rowland, University of California, Irvine, E. S. Saracheck, University of Washington, Seattle, J. M. Wallace, University of California, Seattle.

12. Published by 'Redefining Progress', Anse Miller, Campaign Manager, released 4 April 2002.
13. Lomborg's book was quickly reviewed by a number of journals and daily papers, but most of the reviews were superficial. Obviously the reviewers had seldom gone back to the references that were provided (there are altogether 2930 in the book as a whole, though many were media reports and newspaper articles, and fewer were publications in key scientific journals). A few examples of praising reviews were:

> 'Bjorn Lomborg raises the important question whether the costs of remedying the damage caused by environmental pollution are higher than the costs of the pollution itself. He has written a pioneering book. *Professor R. Rosecrance*, Department of Political Science, UCLA, USA. This had, of course, been done before and had also been dealt with in IPCC (2001c, d).
> 'Lomborg questions most of our common views on the environment, the global food situation, and strategies for development assistance to the poor. He may not be right on all issues, but his plea for scientific stringency, and his exposure of false environmental prophets, are all very credible.' *Stein W Bie*, Director General, International Service for National Agricultural Research (ISNAR).
> 'Based on facts and figures that are common ground to all sides of the ecological debate, this book will change forever the way we think about the state of the world. It is a remarkable, no, an extraordinary achievement.' *Toger Seidenfaden*, Executive Editor-in-chief, Politiken (Copenhagen, Denmark).

14. Academician Kiril Kondratiev has made important scientific contributions in the field of radiation physics and has published a well-known advanced text book on this topic.
15. See Bolin (2004). I submitted my presentation for publication in the proceedings from the conference. I do not know if it was ever published in Russian, but the English version is almost identical to the publication just cited.
16. See Azar and Schneider (2002) and also Section 13.6.
17. The views of the Academy were expressed in 'Declaration of the Advisory Council of the Russian Academy of Sciences on Possible Anthropogenic Climate Change and the Problem of the Kyoto Protocol,' adopted by the Academy on 14 May 2004, signed by Yuri Izrael and available from the Academy's archives.
18. Personal communication by the UK head delegate to the IPCC, David Warrilow; see also *Nature*, editorial 2004.
19. Presentation by James M. Inhofe to the Committee on Environment and Public Works, US Congress, 15 November 2005.

12. Key scientific findings of prime political relevance

1. See R. McKitrick (coordinator) (2007).
2. IPCC (2001a, 2007a).
3. See IPCC (2007a), Jones and Palutikof (2006), Fu *et al.* (2004), Fu *et al.* (2006).
4. A detailed account of the scientific procedure used in order to calibrate climate models against data from the last century is given in IPCC (2001a), Chapters 7, 8 and 9.
5. The role of human-induced aerosols is not yet quantitatively well understood; see e.g. Anderson *et al.* (2003).
6. See also Hansen *et al.* (2005).
7. See IPCC (2001a, 2007a), and also Dore and Mohammed (2005).
8. See ACIA (2004) and Oerlemans (2005).
9. See, Emanuel (2005), Trenberth (2005), Webster *et al.* (2005), Pielke *et al.* (2006), Anthes *et al.* (2006), Curry *et al.* (2006).
10. See Stott *et al.* (2004), Schär *et al.* (2004).
11. See Fu *et al.* (2006).
12. See Royal Meteorological Society (2006).

13. The producers were Michael Stenberg, Linus Torell and Johan Söderberg.
14. The IGBP was initiated in 1986 and is a worldwide programme for the global coordination of research efforts concerning geospheric and biospheric interactions. Its secretariat is located at the Royal Academy of Sciences in Stockholm, Sweden.

13. Climate change and a future sustainable global energy supply

1. Carbon amounts given in gigatons (10^9 tons) of carbon are equivalents of 3.67 Gt of carbon dioxide.
2. See IPCC (2000a), Houghton (1999), Field and Raupach (2004). Compare also with Sections 2.2 and 8.3.
3. See IPCC (2000a).
4. See IPCC (1995a) and Section 8.3.
5. An up-to-date description of the carbon cycle is given in Prentice *et al.* (2001), Field and Raupach (2004).
6. See Cramer *et al.* (2001).
7. See Bolin and Kheshgi (2001) and Baumert *et al.* (2005). The IPCC AR4 contains a modification of Figure 13.1 in that it also includes the net emissions due to deforestation and land-use change. See IPCC (2007c).
8. See also International Energy Agency (2004).
9. See British Petroleum (2004).
10. See Bolin and Kheshgi (2001).
11. As mentioned before, the EU has carefully analysed a scheme for burden sharing between the European countries in order to take into account the differences in the member states' abilities to reduce emissions.
12. See Section 11.1.
13. One may indeed wonder how such a decrease in the number of people in the world will come about, particularly in light of the steadily increasing life expectancy of the world population that is associated with improved standards of living. See Lutz *et al.* (2003).
14. Data from IPCC (1996b) and UNDP (2004); extrapolated to 2005. Reference is also made to International Energy Agency, 2004. Estimated reserves are known and can be exploited, estimated resources remain to be discovered and developed as reserves. The sizes of reserves and resources are dependent on the technology that is or may become available for their exploitation at 'reasonable' costs. Resources may therefore increase with time, when other means of providing energy to society become more costly. However, this does not seem very likely to be the case for conventional oil, the reserves of which have steadily been decreasing The possibility that a rather rapid transformation of the present global energy supply system may be required because of the dwindling resources of conventional oil has been highlighted by The Association for the Study of Peak Oil and Gas APSO (2005).

    In 2007 a critical appraisal of the estimates of reserves and resources of coal was presented (Energy Watch Group, 2007), which indicates that earlier assessments are very uncertain and that resources have probably been considerably overestimated.

References

ACIA, 2004. *Impacts of a Warming Arctic.* New York: Cambridge University Press.

Agarwal, A. and S. Nasrain, 1998. *The Atmospheric right of all people on earth.* Centre on Science and Environment, New Dehli, India.

Agassiz, L., 1840. *Etudes sur les Glaciers* (*Studies on Glaciers*, 1967, translated and edited by A. V. Carozzi. New York: Hafner).

Agrawala, S., 1998. Context and early origins of the Intergovernmental Panel on Climate Change. *Climate Change* **39**, 605–620; Structural and process history of the Intergovernmental Panel on Climate Change. *Climate Change*, **39**, 621–642.

American Meteorological Society, 1996. Open letter to Ben Santer. *Bull. Am. Met. Soc.*, **F77**, 1961–1966.

Anderson, T. L., R. J. Charlson, S. E. Schwartz, *et al.*, 2003. Climate forcing by aerosols – a hazy picture. *Science*, **300**, 1102–1103.

Anthes, R. A., R. W. Corell, G. Holland, J. W. Hurrelll, M-.C. MacCracken, and K. E. Trenberth, 1996. Hurricanes and global warming – potential linkages and consequences. *Bull. Am. Met. Soc.*, **87**, 623–628.

APSO, 2005. Newsletter, 60 – December 2005. APSO, Ireland at www.peakoil.ie

Arrhenius, S., 1896a. On the influence of carbonic acid in the air upon the temperature of the ground. *Phil. Mag.*, **41**, 237–276.

1896b. Nature's heat usage, *Nordisk Tidskrift*, **14**, 121–130 (in Swedish).

Ausubel, J., 1983. Historical note. In *Changing Climate*, Report of the Carbon Dioxide Assessment Committee, pp. 488–491. Washington, DC: National Academy Press.

Azar, C. and S. Schneider, 2002. Are the economic costs of stabilising the atmosphere prohibitive? *Ecological Econ.*, **42**, 73–80.

Azar, C., K. Lindgren, E. Larson, and K. Möllersten, 2006. Carbon capture and storage from fossil fuels and biomass – costs and potential role in stabilizing the atmosphere, *Climatic Change*, **74**, 47–79.

Baumert, K. A., T. Herzog, and J. Pershing, 2005. *Navigating the Numbers, Greenhouse Gas Data and International Climate Policy.* Washington, DC: World Resource Institute.

Benedick, R. E., 1991. *Ozone Diplomacy. New Directions in Safeguarding the Planet.* Cambridge, MA: Harvard University Press.

Bengtsson, L., 2006a. The global energy problem. *Energy & Environment*, **17**, 755–765. 2006b. Geo-engineering; to confine climate change: Is it at all feasible? An editorial comment. *Climate Change*, **77**, 229–234.

262

Black, J., 1754. De homore acido, et magnesia alba. Ph.D. thesis. University of Edinburgh.

Boehmer-Christiansen, S., 1994a. Global climate protection policy: the limits of scientific advice. *Global Environ. Change*, **4**. (a). Part 1, pp. 140–159. (b). Part 2, pp. 185–200.

1994b. A scientific agenda for climate policy. *Nature*, **372**, 400–402.

Bolin, B., 1960. On the exchange of carbon dioxide between the atmosphere and the sea. *Tellus*, **12**, 274–281.

1976. *Energy and Climate*. Stockholm: The Secretariat for Future Studies. Swedish Government.

1977. Changes of land biota and their importance for the carbon cycle. *Science*, **196**, 4290, 613–615.

1979. Global ecology and man. In *Proceedings of the World Climate Conference*. WMO Publ. 537, pp. 27–50. Geneva: WMO.

1980. *Climate changes and their effects on the biosphere*. Fourth IMO Lecture, WMO Publ. 542, Geneva: WMO.

1986. The objectives of the global weather experiment and an overall view of the accomplishments of GARP. In *GARP Publication Series*, No. 26, pp. 3–18. Geneva: WMO.

1991. The global warming issue; scientific knowledge and the challenge for action. In the 1991 Proceedings of the Conference of the International Gas Union. Zürich, Switzerland: International Gas Union.

1992. *The Issue of Global Warming – Knowledge, Uncertainties and the Need for Action*. Proceedings of the Thirteenth World Petroleum Congress, pp. 357–365. Chichester, UK: John Wiley & Sons.

1993a. A joint scientific and political process for a convention on climate change. In G. Sjöstedt and U. Svedin (eds.), *International Environmental Negotiations*, pp. 155–163. Stockholm: Swedish Council for Planning and Coordination of Research & The Swedish Institute for International Affairs.

1993b. Global climate change, an issue of risk assessment and management. NWO Huygens Lecture 1993. www.nwo.nl/nwohome.nsf/Pages/NWOP_6C3L77

1994a. Science and Policy, *Ambio*, **23**, 1, 25–29.

1994b. The interplay of science and politics in the Framework Convention for Climate Change. Opening statement at the Workshop in Fortaleza, Brazil, 17–21 October, 1994. Geneva: IPCC Secretariat, WMO.

1994c. Next step for climate change analysis. *Nature*, **368**, 94.

1995a. Comment on 'Atmospheric residence time and the carbon cycle' by C. Starr. *Energy – The International Journal*, **18**, 1297–1310 (1993). *Energy*, **40**, No. 6, 589.

1995b. Statement to the first session of the Conference of the Parties to the UN Framework Convention on Climate Change in Berlin, 28 March 1995. IPCC Secretariat, WMO, Geneva. Available through Framework Convention on Climate Change, Bonn, Germany.

1996. Climate Change. In *Round up. Energy for our Common World – What will the Future Ask of Us*. World Energy Council, sixteenth Congress. Tokyo, pp. 367–378. London: World Energy Council.

1997. Scientific Assessment of Climate Change. In Ferman, G. (ed.): *International Politics of Climate Change*, pp 83–109. Oslo: Scandinavian University Press,

1998. The Kyoto negotiations on climate change: a scientific perspective. *Science*, **279**, 330–331.

2002. Politics and the IPCC. *Science*, **296**, 1235.

2004. Responding to Climate Change. Knowledge and insight required to act under uncertainty. Focussing on robust findings. SEI Climate and Energy Programme Report 2004–01. Stockholm: Stockholm Environmental Institute.

Bolin, B. and E. Eriksson, 1959. Changes in the carbon dioxide content of the atmosphere and the sea due to fossil fuel combustion. In Bolin, B. (editor) *The Atmosphere and the Sea in Motion*. The Rossby Memorial Volume, pp. 130–142. New York: The Rockefeller Institute Press.

Bolin, B. and J. Houghton, 1995. Berlin and global warming policy. *Nature*, **375**, 176.

Bolin, B. and H. Kheshgi, 2001. On strategies for reducing greenhouse gas emissions. *Proc. Nat. Acad. Sci.*, **98** (9), 4850–4854.

Bolin, B., E. T. Degens, S. Kempe, and P. Ketner, 1979. *The Global Carbon Cycle*, SCOPE 13. Chichester, UK: John Wiley & Sons.

Bolin, B., B. R. Döös, J. Jäger, and R. A. Warrick, 1986. *The Greenhouse Effect, Climate Change, and Ecosystems*. SCOPE 29. Chichester, UK: John Wiley & Sons.

Bolin, B., R. Sukumar, P. Ciais, *et al.* 2000. *Land Use, Land Use Change, and Forestry. Global Perspective*, pp. 25–51. Cambridge: Cambridge University Press.

Boumert, K. E., T. Herzog, and J. Pershing, 2005. *Navigating the Numbers. Greenhouse Gas Data and International Policy*. World Resource Institute Report. Washington, DC: World Resource Institute.

British Petroleum, 2004. The Statistical Review of World Energy.

Bromley, D. A., 1990. The making of a greenhouse policy. *Issues in Science and Technology*, National Academy of Sciences, Fall issue, 55–61.

Budyko, M. and Yu. Izrael, 1987. *Anthropogenic Climate Changes*. Leningrad: L. Gidrometeozat.

Callendar, G. S., 1938. The artificial production of carbon dioxide and its influence on temperature. *QJR. Meteorol. Soc.*, **64**, 223–240.

Charlson, R. J., J. Lagner, H. Rodhe, C. B. Leovy, and S. G. Warren, 1991. Perturbation of the Northern Hemisphere radiative balance by backscattering from anthropogenic sulphate aerosols. *Tellus*, **43A**, 152–163.

Charney, J. 1975. Dynamics of deserts and droughts in Sahel. In *The Physical Basis of Climate and Climate Modelling*, pp. 171–176. Geneva: WMO.

Christy, J. R., R. W., Spencer, and R. T. McNider, 1995. Reducing noise in th MSU daily lower troposphere global temperature data set. *J. Climate*, **8**, 888–896.

Christy, J. R., R. W. Spencer, and W. D. Braswell, 2000. MSU tropospheric temperatures: Data set construction and radiosonde comparisons. *J. Atmos. Oceanic Tech.*, **17**, 1153–1170.

Cline, W. R., 1992. *The Economics of Global Warmning*. Washington, DC: Institute of International Economics.

Coelho, S. T., J. Goldenberg, O. Lucon, and P. Guardabassi, 2005. Brazilian sugarcane ethanol: Lessons learned. Presented at Workshop on Liquid Biofuels, Delhi, Sept. 2005 (sma.goldenberg@cetesb.sp.gov.br).

Cortez, L., 2006. Scaling up the Ethanol Production in Brazil. Opportunities and Challenges. Presentation to the Swedish Energy Agency (STEM) on 6 June 2006.

Council of Environmental Quality, 1981. *Global Energy Futures and the Carbon Dioxide Problem*. Washington, DC: Executive Office of the President.

Cramer, W., A. Bondeau, and F. I. Woodward, 2001. Global response of terrestrial ecosystem structure and function to carbon dioxide and climate change. *Global Change Biology*, **7**, 357–373.

Crawford, E., 1996. *From Ionic Theory to the Greenhouse Effect*. Canton, MA: Science History Publications.

1997. Arrhenius' 1896 model of the greenhouse effect in context. *Ambio*, **26**, 6–11.

Crowley, T. J., 2000. Causes of climate change over the past 1000 years. *Science*, **289**, 270–277.

Crutzen, P. J., 2002. Geology and mankind. *Science*, **295**, 23.

2006. Albedo enhancement by stratospheric sulfur injections: A contribution to resolve a policy dilemma. *Climate Change*, **77**, 211–219.

Crutzen, P., A. R. Mosier, K. A. Smith, and W. Winiwarter, 2007. N_2O release from agro-biofuel production negates climate effect of fossil fuel CO_2 'savings'. *Nature* (in press).

Curry, J. A., P. J. Webster, and G. J. Holland, 2006. Mixing politics and science in testing the hypothesis that greenhouse warming is causing a global increase in hurricane intensity. *Bull. Amer. Met. Soc.*, **87**, 1025–1037.

de Moor, A. P. G., M. M. Bark, M. G. J. den Elzen, and D. P. van Vuuren, 2002: Evaluating the Bush climate change initiative. RIVM Report 728001019, Dutch Ministry of the Environment.

Dore, Mohammed, H. I., 2005. Climate Change and changes in global precipitation patterns. What do we know? *Environment Intern.*, **31** (October), 1167–1181.

Emanuel, K. A., 2005. Increasing destructiveness of tropical cyclones over the last 30 years. *Nature*, **436**, 686–688.

Energy Watch Group, 2007. Coal: resources and future production. Paper 1/07. www.energywatchgroup.org

Enting, I., T. M. Wigley, and M. Heimann, 1994. Future emissions and concentrations of carbon dioxide: Key atmosphere/ocean/land analyses. CSIRO Division of atmospheric research. Technical Paper, No 31.

Environment Canada, WMO, UNEP, 1988. *The Changing Atmosphere*, Conference Proceedings. No. 710. Toronto: WMO/OMM.

Fan, S. M., M. Gloor, J. Mahlman, *et al.* 1998. A large terrestrial sink in North America implied by atmospheric and oceanic carbon dioxide data and models. *Science*, **282**, 442–446.

Field, B. and M. R. Raupach, 2004. *The Global Carbon Cycle. Integrating Humans, Climate and the Natural World*. Washington, DC: Island Press.

Fourier, J., 1824. *Mémoirs de l'Académie royale des science de l'Institut de France*, pp. 585–587. Paris: Didot Père et Fils.

From, E. and C. D. Keeling, 1986. Reassessment of late nineteenth century atmospheric carbon dioxide variations in the air of Western Europe and the British Isles based on an unpublished analysis of contemporary air masses by G. S. D. Callendar, *Tellus* **38B**, 87–105.

Fu, Q. and C. M. Johanson, 2004. Contribution of stratospheric cooling to satellite-inferred tropospheric temperature trends. *Nature*, **429**, 55–58.

Fu, Q., C. M. Johanson, J. M. Wallace, and T. Reichler, 2006. Enhanced mid-latitude tropospheric warming in satellite measurements. *Science*, **312**, 1179.

Gore, A., 1992. *Earth in Balance; Ecology and the Human Spirit*. Boston, MA: Houghton Mifflin.

Hansen, J., 1988. The greenhouse effect: impacts on current global temperature and regional heat waves. Testimony before the Committee on Energy and Natural Resources, US Senate, Washington DC, June 23.

Hansen, J., L. Nazarenko, R. Ruedy, *et al.*, 2005. Earth's energy imbalance; confirmation and implications, *Science*, **308**, 1431–1435.

Harrison, R. G. and Stephenson, 2006. Empirical evidence for a nonlinear effect of galactic cosmic rays on clouds. *Proc. R. Soc. A*, **462** (2068), 1221–1233.

Hasselmann, K., 1997a. Are we seeing global warming? *Science*, **276**, 914–915.
    1997b. Eine Antwort auf die Kritik an der Klimaforschung. *Die Zeit*, **31**.
Hecht, A. D. and D. Tirpak, 1994. Framework agreement on climate change: a scientific and policy history. *Climate Change*, **29**, 371–402.
Heimann, M., 1997. A review of the contemporary global carbon cycle as seen a century ago by Arrhenius and Högbom. *Ambio*, **26**, 17–24.
Houghton, J., 2004. *Global Warming. The complete briefing*, Third edition. Cambridge: Cambridge University Press.
Houghton, R. A., 1999. The annual flux of carbon to the atmosphere from changes in land use 1850–1990. *Tellus*, **51B**, 298–313.
Houghton, R. A., J. E. Hobby, J. M. Melillo, *et al.*, 1983. Changes in the carbon content of terrestrial biota and soils between 1860 and 1980. A net release of carbon dioxide to the atmosphere. *Ecolog. Monogr.*, **53**, 235–262.
Hurrell, J. W. and K. E. Trenberth, 1997. Spurious trends in satellite MSU temperatures from merging different satellite records. *Nature*, **86**, 164–167.
ICSU/IUGG, CAS, and COSPAR, 1967. Report on the Study Conference on The Global Atmospheric Research Programme, held in Stockholm, 70 pp and 8 appendices, including a Report of COSPAR Working Group VI. Geneva: WMO.
IGBP Terrestrial Carbon Working Group, 1998. The terrestrial carbon cycle: implications for the Kyoto protocol. *Science*, **280**, 1393–1394.
International Energy Agency, 1997. *CO_2 Emissions from Fuel Combustion*. Paris: OECD.
    2004. Key World Energy Statistics. Paris: OECD.
    2006. Outlook 2006. EIA-0484. Paris: OECD.
IPCC, 1988. Report of the First Session of the WMO/UNEP Intergovernmental Panel on Climate Change, Geneva, November 1988. Geneva: WMO.
    1989a. Report of the First Session IPCC, Working Group III February 1989. Geneva: WMO.
    1989b. Report of the Second Session of the WMO/UNEP Intergovernmental Panel on Climate Change, Nairobi, June 1989. World Climate Programme Publications Series. Geneva: WMO.
    1990a. *Climate Change, The IPCC Scientific Assessment*. Edited by J. T. Houghton, G. J. Jenkins, and J. J. Ephraim, Working Group I. Cambridge: Cambridge University Press.
    1990b. *Climate Change, The IPCC Impact Assessment*. Edited by W. J. McG. Tegart, G. W. Sheldon, and D. C. Griffiths, Working Group II. Canberra: Australian Government Publishing Service.
    1990c. *Climate Change. The IPCC Response Strategies*. Working Group III, coordinator F. M. Bernthal. Washington, DC: US National Science Foundation.
    1990d. *Report of the Third Session of the WMO/UNEP Intergovernmental Panel on Climate Change, Washington, DC, February, 1990*, IPCC-5. Geneva: WMO.
    1990e. *Report of the Fourth Session of the WMO/UNEP Intergovernmental Panel on Climate Change, Sundsvall, Sweden, August, 1990*. IPCC-6. Geneva: WMO.
    1991a. *Report of the Fifth Session of the WMO/UNEP Intergovernmental Panel on Climate Change, Geneva, March 1991*. IPCC-7. Geneva: WMO.
    1991b. *Report of the Sixth Session of the WMO/UNEP Intergovernmental Panel on Climate Change, Geneva, October 1991*. IPCC-8. Geneva: WMO.
    1992a. Climate Change, *The IPCC 1990 and 1992 Assessments. IPCC First Assessment Report*. Overview and Policymakers' Summaries, and 1992 IPCC Supplement. Geneva: WMO and UNEP.

1992b. *Climate Change 1992, The Supplementary Report to the IPCC Scientific Assessment*. Cambridge: Cambridge University Press.

1992c. *Climate Change 1992, The Supplementary Report to the IPCC Impacts Assessment*. Canberra: Australian Government Publishing Service.

1992d. *Report of the Eighth Session of the WMO/UNEP Intergovernmental Panel on Climate Change. Harare, November 1992*. Geneva: WMO.

1993. *Report of the Ninth Session of the WMO/UNEP Intergovernmental Panel on Climate Change, Geneva, June 1993*. Geneva: WMO.

1994a. *Report of the Fourth Plenary Session of the Scientific Assessment Working Group (WGI) of the IPCC. Maastricht, The Netherlands, 13–15 September 2004*. Geneva: WMO.

1994b. *Report of the Tenth Session of the WMO/UNEP Intergovernmental Panel on Climate Change, Nairobi, November 1994*. Geneva: WMO.

1995a. *Climate Change 1994. Radiative forcing and climate change and an evaluation of the IS92 Emissions Scenarios*. Cambridge: Cambridge University Press.

1995b. *Report of the Eleventh Session of the WMO/UNEP Intergovernmental Panel on Climate Change, Rome December 1995*. Geneva: WMO.

1996a. *Climate Change 1995, The Science of Climate Change*. Contributions by Working Group I to the Second IPCC Assessment Report. Cambridge: Cambridge University Press.

1996b. *Climate Change 1995, Impacts, Adaptations and Mitigation of Climate Change. Scientific and Technical Analyses*. Contributions by Working Group II to the Second IPCC Assessment Report. Cambridge: Cambridge University Press.

1996c. *Climate Change 1995, Economic and Social Dimensions of Climate Change*. Contributions by Working Group III to the Second IPCC Assessment Report. Cambridge: Cambridge University Press.

1996d. *Climate Change 1995. IPCC Second Assessment. A Synthesis Report*. Geneva: WMO.

1996e. Chairman's report to the second session of the SBSTA. 27 February 1996, In *Report of the Second SBSTA Session*. Bonn: FCCC.

1996f: Technical Paper 1, *Technologies, Policies and Measures for Mitigating Climate Change*. Geneva: WMO.

1996g. Chairman's report to the fourth session of the SBSTA, 16 December, 1996. In *Report of the Fourth SBSTA Session*. Bonn: FCCC.

1996h. *Report on the Twelfth Session of the UNEP/WMO Intergovernmental Panel on Climate Change. Mexico City, 11–13 September 1996*. Geneva: WMO.

1997a. Technical Paper 2, *An Introduction to Simple Climate Models Used in the IPCC Second Assessment Report*. Geneva: WMO.

1997b. Technical Paper 3, *Stabilization of Atmospheric Greenhouse Gases: Physical, Biological and Socio-economic Implications*. Geneva: WMO.

1997c. Technical Paper 4, *Implications of Proposed CO_2 Emissions Limitations*. Geneva: WMO.

1997d. Chairman's report to the fifth session of SBSTA, 25 February, 1997. In *Report of the Fifth SBSTA Session*. Bonn: FCCC.

1997e. *Report on the Thirteenth Session of the UNEP/WMO Panel on Climate Change in the Maldives. 22–28 September 1997*. Geneva: WMO

1997f. Chairman's report to the seventh session of SBSTA, 24 October, 1997. In *Report of the Seventh SBSTA Session*. Bonn: FCCC.

1998. *The Regional Impact of Climate Change, an Assessment of Vulnerability.* Cambridge: Cambridge University Press.

1999. *Aviation and the Global Atmosphere.* Cambridge: Cambridge University Press.

2000a. *Land Use, Land-Use Change, and Forestry.* Cambridge: Cambridge University Press.

2000b. *Special Report on Emissions Scenarios.* Cambridge: Cambridge University Press.

2000c. *Methodological and Technological Issues in Technology Transfer.* Cambridge: Cambrdige University Press.

2001a. *Climate Change 2001. The Scientific Basis.* Cambridge: Cambridge University Press.

2001b. *Climate Change 2001. Impacts, Adaptation and Vulnerability.* Cambridge: Cambridge University Press.

2001c. *Climate Change 2001. Mitigation.* Cambridge: Cambridge University Press.

2001d. *Climate Change 2001. Synthesis Report.* Cambridge: Cambridge University Press.

2005. *Special Report on Carbon Dioxide Capture and Storage, Summary for Policymakers and Technical Summary.* Geneva: WMO.

2007a. *Climate Change 2007: The Physical Science Basis.* Contribution of Working Group I to the Fourth Assessment Report of the Intergovernmental Panel on Climate Change, S. Solomon, D. Qin, M. Manning *et al.* (eds). Cambridge: Cambridge University Press.

2007b. *Climate Change 2007: Climate Change Impacts, Adaptations and Vulnerability.* Contribution of Working Group II to the Fourth Assessment Report of the Intergovernmental Panel on Climate Change. Cambridge: Cambridge University Press.

2007c. *Climate Change 2007: Mitigation.* Contribution of Working Group III to the Fourth Assessment Report of the Intergovernmental Panel on Climate Change. Cambridge: Cambridge University Press.

IPCC/TEAP, 2005. *Safeguarding the Ozone Layer and the Global Climate System.* Cambridge: Cambridge University Press.

JOC, 1968. *Report of the 1st JOC Session, (April 1968), Geneva.* Geneva: WMO.

1971. *Report of the 6th JOC Session (September 1971), Toronto.* Geneva: WMO.

1972. *Report of the 7th JOC Session (June/July 1972), Munich.* Geneva: WMO.

1973. *Report of the 8th JOC Session (March 1973), London.* Geneva: WMO.

1975. *The Physical Basis of Climate and Climate Modelling*, Report of the International Study Conference in Stockholm, 29 July–10 August 1974. GARP Publications Series, No 16. Geneva: WMO.

1979. *Report on the JOC Study Conference on Climate Models: Performance, Inter-comparison and Sensitivity Studies*, GARP Publication Series, No. 22. Geneva: WMO.

1986. *International Conference on the Results of the Global Weather Experiment and Their Implications for the World Weather Watch.* GARP Publications Series, No 26. Geneva: WMO.

Keeling, C. D., 1958. The concentration and isotope abundances of atmospheric carbon dioxide in rural areas. *Geochem. Cosmochem. Acta*, **13**, 322–334.

1960. The concentration and isotopic abundances of carbon dioxide in the atmosphere. *Tellus*, **12**, 200–203.

Keeling, R. F., S. P. Naijar, and M. Heiman, M., 1996. Global and hemispheric CO_2 sinks deduced from changes in atmospheric O_2 concentration. *Nature*, **381**, 218–221.

Kostitzin, V. A., 1935. *Evolution de l'atmosphère, circulation organique, époques glacières*. Paris: Hermann.

Langley, S., 1889. The temperature of the moon. *Mem. Nat. Acad. Sci.*, **4**, 107–212.

Leggett, J. (ed.), 1990. *Global Warming; The Greenpeace Report*. Oxford: Oxford University Press. 554 pp.

Linden, H. R., 1993. A dissenting view on global climate change. *The Electricity Journal*, July, 62–69.

Lindzen, R. S., 1990. Some coolness concerning global warming. *Bull. Amer. Met. Soc.*, **71**, 288–299.

1997. Can increasing atmospheric CO_2 affect global climate? *Proc. Natl. Acad. Sci., USA* **94**, 8335–8342.

Lindzen, R. S., S. Ming-Dah, and A. Y. Hou, 2001. Does the earth have an adaptive infrared iris? *Bull. Am. Met. Soc.*, **82**, 417–432.

Lomborg, B., 2001. *The Skeptical Environmentalist*. Cambridge: Cambridge University Press. pp 515.

2002. The sceptical environmentalist, replies. *Sci. Amer.*, May, 9–10.

Lorenz, E. N., 1963. Deterministic non-periodic flow. *J. Atmos. Sci.*, 130–141.

1975. Climatic Predictability. In *The Physical Basis of Climate and Climate Modelling, JOC (1975)*, pp. 132–136. Geneva: WMO.

Lotka, A. J., 1924. *Elements of Physical Biology*, Baltimore, MD: Williams & Wilkins.

Lutz, W., B. C. F. O'Neill, and S. Scherbov, 2003. Europa's population at a turning point. *Science*, **299**, 1991–1992.

MacCracken, M., A. D. Hecht, M. Budyko and Yu. Izrael, 1990. A Special US–USSR Report on Climate and Climate Change. Washington, DC: Lewis Press.

Malone, T. F. and J. G. Roederer (eds.), 1985. *Global Change, The Proceedings of a Symposium by the International Council of Scientific Unions (ICSU) during the Twentieth General Assembly in Ottawa, Canada, September, 1994*. Cambridge: Cambridge University Press.

Manabe, S., 1975. The use of comprehensive general circulation modelling for studies of the climate and climate variations. In JOC, *The Physical Basis of Climate and Climate Modelling*, 1975, pp. 148–162. Geneva: WMO.

Manabe, S. and R. T. Wetherald, 1967. Thermal equilibrium of the atmosphere via given distribution of relative humidity. *J. Atmos. Sci.*, **24**, 241–259.

Mann, M. E., J. Park, and R. S. Bradley, 1995. Global interdecadal and century scale climate oscillations during the past five centuries. *Nature*, **378**, 266–270.

Mann, M. E., R. S. Bradley, and M. K. Hughes, 1999. Northern hemisphere temperatures during the past millennium, inferences, uncertainties and limitations. *Geophys. Res. Letters*, **26**, 759–762.

Marsh, N. D., and H. Svensmark, 2003. Solar influence on earth's climate. *Space Sci. Rev.*, **107**, 317–325.

McIntyre, S. and R. McKitrick, 2003. Correction to the Mann *et al.*, (1998) proxy data base and Northern Hemisphere temperature series. *Energy & Environment*, **14.6**, 751–771.

McKitrick, R. (coordinator), 2007. *Independent Summary for Policymakers for the IPCC Fourth Assessment Report*. Vancouver, BC: The Fraser Institute.

Meadows, D. H., J. Randers, and W. W. Beherens III, 1972. *The Limits of Growth. A report from the Club of Rome's Project on the Predicament of Mankind*. New York: Universe Books.

Mintzer, M. and J. A. Q. Leonard (eds), 1994. *Negotiating Climate Change, The Inside Story of the Rio Convention*. Cambridge: Cambridge University Press.

National Academy of Sciences, 1966. *The Feasibility of a Global Observation and Analysis Experiment*, Publication 1290, Washington, DC: NAS.

1975. *Understanding Climate Change, A Program for Action*. US Committee for GARP. Washington, DC: NAS.

1977. *Energy and Climate*. Geophysics Study Committee, Washington, DC: NAS.

1979. *Carbon Dioxide and Climate, A Scientific Assessment*. US Climate Research Board. Washington, DC: NAS.

1982. *Carbon Dioxide and Climate: A Second Assessment*. Washington, DC: NAS.

1983. *Changing Climate*. Report of the Carbon Dioxide Assessment Committee. Washington, DC: NAS.

National Research Council, 2001. *Climate Change Science, An Analysis of Some Key Questions*. Washington, DC: National Academy Press.

2004a. *Surface Temperature Reconstructions for the Last 2000 Years*. Washington, DC: National Academy Press.

2004b. The Hydrogen Economy. Opportunities, Costs, Barriers, and R&D Needs. Washington, DC: National Academy Press.

Nature, Editorial, 1995. *Nature*, **375,** 176.

Nature, Editorial, 2004. Crunch time for Kyoto, *Nature*, **431,** 12–13.

Nilsson, S., M. Jonas, V. Stolbovoi, A. Shvidenko, M. Obersteiner, and MacCallum, 2003. The missing 'missing sink'. *The Forestry Chronicle*, **79**.6, 1071–1074.

Nordhaus, W. D., 1991. To slow or not to slow. The Economics of the Greenhouse Effect. *Economic Journal*, **101,** No. 6, 920–937.

Nordhaus, W. D. and G. W. Yohe, 1983. Future carbon emissions from fossil fuel. In *Changing Climate*. Washington, DC: National Research Council. National Academy Press.

Oerlemans, J., 2005. Extracting a climate signal from 169 glacier records. *Science*, **308,** 675–677.

Pacala, S., and R. Socolow, 2004. Stabilization wedges: solving the climate problem for the next 50 years with current technologies. *Science*, **305,** 968–971.

Patzek, T. W., and O. Piementel, 2007. Thermodynamics of energy production from biomass. *Critical Revs. in Plant Sci.* (in press).

Pepper, W., J. U. Leggett, R. Swart, J. Wasson, J. Edmonds, and I. Mintzer, 1992. *Emissions Scenarios for the IPCC, An Update, Assumptions, Methodology, and Results*. Prepared for IPCC Working Group I.

Pielke, R., Ch. Landsea, M. Mayfield, J. Lavers, and T. Pasch, 2006. Hurricanes and global warming. *Bull. Am. Met. Soc.* **86,** 1571–1575.

Pollách, H. N., S. Huang, and P. Y. Shen, 1998. Climate change record in sub-surface temperatures: A global perspective. *Science*, **282,** 279–281.

Pouillet, C., 1837. Mémoire sur la chaleur solaire, sur les pouvoirs rayonnonants et absorbants de l'air atmosphérique, et sur la température de l'espace. *Comptes rendus hebdomadaires des séances de l'Académie des Sciences*, 7, 24–65.

Prentice, C., *et al.*, 2001. The carbon global cycle and atmospheric carbon dioxide. In *Climate Change 2001. The Scientific Basis*. pp. 183–237. Cambridge: Cambridge University Press.

Ramanathan, V., 1975. Greenhouse effects due to chlorofluorocarbons: climate implications. *Science*, **190,** 50–52.

Ramanathan, V., R. J. Cicerone, H. B. Singh, and J. T. Kiehl, 1985. Tracer gas trends and their potential role in climate change. *J. Geophys. Res.*, **90,** D3, 5547–5566.

Ramanathan, V. and A. M. Vogelmann, 1997. Greenhouse effect, atmospheric solar absorption and the earth's radiation budget: from the Arrhenius–Langley era to the 1990s. *Ambio* **26**, 38–46.

Revelle, R. and H. Suess, 1957. Carbon dioxide exchange between atmosphere and oceans and the question of an increase of atmospheric carbon dioxide during the past decades. *Tellus*, **9**, 18–27.

Riahi, K., A. Grübler, and N. Nakicenovic, 2006. Scenarios of long-term socio-economic and environmental development under climate stabilization. *Technological Forecasting and Social Change* (to be published).

Rodhe, H., R. Charlson, and E. Crawford, 1997. Svante Arrhenius and the greenhouse effect. *Ambio*, **26**, 2–5.

Royal Meterological Society, 2006. Special Issue. Climate change in high latitudes. 1 & 2. *Weather*, **61**, No. 3 and 4.

Santer, B. D., K. E. Taylor, T. M. L. Wigley, *et al.*, 1996. A search for human influences on the thermal structure of the atmosphere. *Nature*, **382**, 39–46.

SCEP, 1970. *Man's Impact on the Global Environment*. Cambridge, MA: MIT Press.

Schär, Ch., P. L. Vidale, D. Lüthi, *et al.*, 2004. The role of increasing temperature variability in European summer heat waves. *Nature*, **427**, 332–336.

Schneider, S., 1976. *The Genesis Strategy*. Plenum Publications.

Schneider, S., J. P. Holdren, J. Bongaards, and T. Lovejoy, 2002. Misleading math about the earth. *Sci. Am.*, January, 2001, 63–73.

Schneider-Carius, K., 1955. *Wetterkunde, Wetterforsch-ung*. Freiburg: Verlag Karl Alber.

Simmons, A. J., P. D. Jones, V. da Costa Bechtold, *et al.*, 2004. Comparison of trends and low-frequency variability in CRU, ERA-40, and NCEP/NCAR analyses of surface air temperature. *J. Geoph. Res.*, **109**, D24115.

Singer, F. (editor), 1992. *The Greenhouse Debate Continued: An Analysis and Critique of the IPCC Climate Assessment*. The Science and Environmental Policy Project, San Francisco, CA: ICS Press.

SMIC, 1971. *Study on Man's Impact on Climate*. Cambridge, MA: MIT Press.

Spencer, R. W. and J. R. Christy, 1993. Precision lower stratospheric temperature monitoring with MSU. Validation and results 1979–91. *J. Climate*, **6**, 1194–1204.

Starr, Ch., 1993. Atmospheric CO_2 residence time and the carbon cycle. *Energy*, **18** No 12, 1297–1310.

Stern, N., 2006. *The Economics of Climate Change*. The Stern review's final report. London: Treasury of UK Government.

Stott, P. A., D. A. Stone, and M. R. Allen, 2004. Human contribution to European heat wave 2003. *Nature*, **432**, 610–613.

Stott, P. A., S. F. B. Tett, J. S. Jones, M. R. Allen, W. J. Ingram, and J. F. B. Mitchell, 2001. Attribution of twentieth century temperature change to natural and anthropogenic causes. *Clim. Dyn.*, **13**, 1–22.

Svensmark, H., 1998. Influence of cosmic rays on earth's climate. *Phys. Rev. Letters*, **81**, 5027–5030.

SWECLIM, 2003. Rossby Centre. See <www.SMHI.se>

Tata Energy Research Institute, TERI, 1989. Statement from the conference 21–23 February, 1989, New Delhi.

Trenberth, K., 2005. Uncertainties in hurricanes and global warming. *Science*, **308**, 1753–1754.

Tyndall, J. 1865. *Heat Considered as a Mode of Motion*, 2nd edition London: Longmans, Green and Co.

UNESCO-SCOPE, 2006. *Policy Briefs*, October, No 2.

UNDP, 2004. *World Energy Assessment, Overview.*

United Nations, 1992. *Framework Convention on Climate Change*. Geneva: UNEP/ WMO, Information Unit on Climate Change.

United States 104th Congress; Committee on Science, 1996. *Environmental Science under Siege: Fringe Science and the 104th Congress.*

Veizer, J., 2005. Celestial climate driver: a perspective from four billion years of the carbon cycle. *Geosci. Canada*, **32**, No 1, 13–28.

Vernadsky, I. V., 1926. *The Biosphere* (in Russian). Published in French in 1929: *La Biosphere*. Paris: Alkan.

Ward, B. and R. Dubos, 1972. *Only One Earth*: *Report on the Human Environment.* W. W. Norton & Company.

Webster P. J., G. J. Holland, J. A. Curry, and H.-R. Chang, 2005. Changes in tropical cyclone number, duration and intensity in a warming environment. *Science*, **309**, 1844–1846.

Wentz, F. J. and M. Schabel, 1998. Effects of orbital decay on satellite derived lower troposphere temperature trends. *Nature*, **394**, 661–664.

Western Fuel Association, 1996. *State of the Climate Report*. 4301 Wilson Boulevard, Suite #805, Arlington, VA 22203-1860.

Wigley, T. M. L., R. Richels, and J. A. Edmonds, 1996. Economic and environmental choices in the stabilisation of atmospheric CO_2 concentrations. *Nature*, **379**, 242–245.

WMO, 1979. *Proceedings of the World Climate Conference*, WMO Publication No 537. Geneva: WMO.

WMO/ICSU/UNEP, 1981. On the assessment of the role of carbon dioxide on climate variations and their impact. Geneva: WMO.

World Commission on Environment and Development, 1987. *Our Common Future*. Oxford: Oxford University Press.

World Energy Council, 1993. *Energy for Tomorrow's World*. London: Kogan Page.

Name index

Agassiz, Louis, 3
Al-Sabban, 113
Al-Gain, Abdulla, 83
Arrhenius, Svante, 4, 5–6

Baker, James, 53
Bjerknes, Vilhelm, 4
Böhmer-Christiansen, Sonja, 100, 188
Bongaarts, 184
Bromley, Allan, 61
Brown, George, 133
Brundtland, Gro Harlem, 43, 48
Budyko, M., 64
Bush, George, 57
Bush, George, W., 178, 180

Callender, G. S., 11
Carter, Jimmy, 34
Charney, Jule, 20, 33
Cicerone, Ralph, 179
Clausen, Eilean, 132, 143
Clinton, William (Bill), 142–143, 190
Croll, James, 4
Cutajar, Michael, 86

de Geer, Gerhard, 4
Dietze, Peter, 136
Dowdeswell, Elisabeth, 70, 83, 87

Emsley, Peter, 136
Estrada-Ouyela, Raul, 85, 86

Flanery, Brian, 143

Garcia, Ronaldo, 24
Gibson, J., 103
Goodman, Gordon, 39
Gore, Al, 153, 212, 225

Hansen, James, 49, 63
Hasselmann, Klaus, 136
Houghton, Sir John, 64, 113

Illiaronov, 187
Inhofe, James, 191
Izrael, Yu, 64, 83, 188

Jefferson, Michael, 92

Keeling, Charles, 8, 34
Kennedy, John F., 19
Kondratiev, Kiril, 187

Lee, Hoesung, 83
Linden, H., 99
Lindzen, Richard, 136, 182, 188
Lomborg, Bjorn, 183
Lorenz, Edward, 30, 89
Lotka, Alfred, James 10

MacNeill, J., 40, 48
Mahlman, Jerry, 134
Malone, Thomas, 22, 39
Manabe, Suki, 28, 29
Meira, Filho Gylvan, 83, 139
Merkel, Angela, 108
Michaels, Patrik, 128, 134

Nierenberg, William, 72

Obasi, Patrick, 84

Pachauri, R., 186
Pearlman, Donald, 85, 92, 103, 130
Putin, Vladimir, 187, 189

Reinstein, Robert, 83
Ripert, Jean, 55, 69, 77

Santer, Benjamin, 112, 129
Schmidt Heine, 98
Schneider, Steven, 33, 184
Seitz, I., 128, 157
Shlaes, John, 85, 92, 103, 130
Singer, Fredrick, 73, 134

Subject index

Printed in the United Kingdom
by Lightning Source UK Ltd.
134110UK00001B/153/P

A HISTORY OF THE SCIENCE AND POLITICS OF CLIMATE CHANGE

The Role of the Intergovernmental Panel on Climate Change

The issue of human-induced global climate change became a major environmental concern during the twentieth century, and is the paramount environmental debate of the twenty-first century. Response to climate change requires effective interaction from the scientific community, society in general, and politicians in particular. The Intergovernmental Panel on Climate Change (IPCC), formed in 1988, has gradually developed to become the key UN body in providing this service to the countries of the world.

Written by its first Chairman, this book is a unique overview of the history of the IPCC. It describes and evaluates the intricate interplay between key factors in the science and politics of climate change, the strategy that has been followed, and the regretfully slow pace in getting to grips with the uncertainties that have prevented earlier action being taken. The book also highlights the emerging conflict between establishing a sustainable global energy system and preventing a serious change in global climate. This text provides researchers and policy makers with an insight into the history of the politics of climate change.

BERT BOLIN is Professor Emeritus in the Department of Meteorology at the University of Stockholm, Sweden. He is a former Director of the International Institute for Meteorology in Stockholm, and former Scientific Advisor to the Swedish Prime Minister. He was Chairman of the IPCC from 1988 to 1997. Professor Bolin has received many awards during his career, including the Blue Planet Prize from the Asahi Glass Foundation, the Rossby Medal from the American Meteorological Society, the Global Environmental Leadership Award from the World Bank, and the Arrhenius Medal from the Royal Swedish Academy of Sciences.